MOLECULAR BIOLOGY OF THE MYOCARDIUM

MOLECULAR BIOLOGY OF THE MYOCARDIUM

Edited by
Michihiko Tada, M.D., Ph.D.
Department of Medicine and Pathophysiology
Osaka University School of Medicine
Suita, Osaka 565, Japan

JAPAN SCIENTIFIC SOCIETIES PRESS
Tokyo

CRC PRESS
Boca Raton Ann Arbor London Tokyo

Supported in part by the Ministry of Education, Science and Culture under Grant-in-Aid for Publication of Scientific Research Result.

Published jointly by:
JAPAN SCIENTIFIC SOCIETIES PRESS
2-10 Hongo, 6-chome, Bunkyo-ku, Tokyo 113, Japan
ISBN 4-7622-5685-4

and

CRC PRESS
2000 Corporate Blvd., N.W., Boca Raton, FL 33431, U.S.A.
ISBN 0-8493-7752-8

Library of Congress Cataloging-in-Publication Data

Molecular biology of the myocardium/edited by Michihiko Tada. p.   cm.
    Includes index.
    ISBN 0-8493-7752-8
    1. Myocardium—Molecular biology. I. Tada. Michihiko, 1938
    [DNLM: 1. Calcium Channels—physiology.   2. Molecular
    Biology.   3. Myocardium—chemistry.   4. Myocardium—cytology.
    5. Myocardium—enzymology.   WG 280 M718]
QP114. M65M65   1992
612, 1'73—dc20
DNLM/DLC
For Library of Congress   92-13557
                                                                        CIP

Distributed in all areas outside Japan and Asia between Pakistan and Korea by CRC Press.

Printed in Japan

Preface

Cardiovascular medicine has been very tardy in applying the techniques and achievements of molecular biology. In many other fields of biomedical research, such application has been proven to be at least somewhat useful to gain insights into key problems for which conventional methods and theories have been considered to be of no use. As we move into the next century, molecular biology in basic and clinical medicine will inevitably become widespread. Cardiovascular researchers and clinicians with high incentive and awareness have certainly recognized this current consensus of views, and have already initiated their own attempts to integrate molecular biological approaches in their research projects. As we have begun to see in the literature, books and at meetings, North American colleagues in cardiovascular medicine have been the earliest to do this, while Europeans and Japanese have hesitated and are late in starting.

This book represents the first attempt in Japan to publish as a monograph a concise documentation of the evolution of achievements in molecular biology in cardiovascular medicine. It is based primarily upon findings resulting from four-year, multi-institutional research project sponsored by the Ministry of Education, Science and Culture of Japan. At the conclusion of this project, the research participants in Japan met with leading scientists from abroad active in related field and held a two-day intensive session in an international atmosphere.

It is clear what is presented here that the molecular biological approaches to cardiovascular medicine can be promising in solving certain problems of physiological and pathophysiological importance, although some achievements of record are still premature or in their elementary stages. With these as stepping stones, we feel that the next decade will see more challenges and advances, leading to more realistic linkages between molecular biology and medicine in cardiovascular research and practice.

October 1991   Michihiko TADA

Contents

CONTENTS

Molecular Biology of Sarcoplasmic Reticulum Proteins Involved in EC Coupling

DAVID M. CLARKE, KEI MARUYAMA, TIP W. LOO, AND
DAVID H. MACLENNAN

*Banting and Best Department of Medical Research, University of Toronto, C.H. Best
Institute, Toronto, Ontario M5G 1L6, Canada*

The sarco(endo)plasmic reticulum of muscle and nonmuscle cells regulates intracellular Ca^{2+} concentrations by pumping Ca^{2+} to the lumen of the reticular membrane through a Ca^{2+}-dependent ATPase, storing it in association with various Ca^{2+} binding proteins and releasing it through a Ca^{2+} release channel. Our goal in early studies of the sarcoplasmic reticulum was to isolate and characterize the major proteins of the system in order to understand how they relate to Ca^{2+} transport, sequestration, and release. With the advent of recombinant DNA technology, we began in the early 1980s to clone cDNAs encoding these proteins. These approaches, carried out in our own and in other laboratories throughout the world, have led to a rather thorough understanding of the composition of the sarco(endo)plasmic reticulum in different tissues.

In fast-twitch skeletal muscle, the Ca^{2+} pump is the product of the ATP2A, localized on human chromosome 16 (*1*). This gene gives rise to two alternatively spliced transcripts which, in turn, give rise to an adult Ca^{2+} pump, designated SERCA1a, and a neonatal pump, designated SERCA1b (*2*). A second gene, ATP2B, located on human chromosome 12 (*1*), also gives rise to two products of alternative splicing, SERCA2a which is expressed in cardiac and slow-twitch muscles, and a longer form, SERCA2b, expressed in smooth muscle and nonmuscle tissues (*3, 4*). Phospholamban, a regulator of the SERCA2 gene product is expressed in cardiac, slow twitch, and smooth muscle tissues (*5*). It is expressed from a single gene located on human chromosome 6 (*6*).

1

Several Ca²⁺ binding proteins exist in the lumen of the sarco(endo)-plasmic reticulum. In fast- and slow-twitch skeletal muscle, the product of the fast-twitch calsequestrin gene (CAQ1) (*7, 8*), located on human chromosome 1 (*9*), is the major Ca²⁺ binding protein. In both skeletal and cardiac muscle, sarcalumenin (*10, 11*) expressed from a gene which is alternatively spliced to form glycoprotein products of apparent mass 160 and 53 kDa, and a histidine rich Ca²⁺ binding protein (*12, 13*) also appear to contribute to luminal Ca²⁺ binding. A second calsequestrin gene (CAQ2), encodes a major luminal Ca²⁺ binding protein in cardiac muscle and a minor Ca²⁺ binding protein in slow-twitch muscle (*14*). Neither of these calsequestrin genes appears to be expressed in nonmuscle tissues where luminal Ca²⁺ binding may be taken over by an analogue of calsequestrin designated calreticulin (*15*). Calreticulin and calsequestrin have similar molecular weights, are highly acidic, bind large amounts of Ca²⁺ with low affinity and are predicted to have structured NH_2-terminal regions and relatively unstructured COOH-terminal sequences in which Ca²⁺ binding sites are located. The calreticulin gene has been localized to human chromosome 19 (*16*).

At least three genes encode Ca²⁺ release channels. The RYR1 gene, localized on human chromosome 19q13.1 (*17*) is expressed almost exclusively in fast- and slow-twitch skeletal muscle (*18*). The RYR2 gene, localized on human chromosome 1, is expressed in cardiac muscle and brain tissue (*18*). The Ca²⁺ release channel function in smooth muscle and nonmuscle tissues appears to be carried out by the $1P_3$ receptor, an analogue of the ryanodine receptor (*19*).

In cardiac muscle, then, the sarcoplasmic reticulum is made up from the SERCA2a gene product, the regulatory protein phospholamban, the RYR2 gene product, and the CAQ2 gene product. In fast-twitch skeletal muscle, the sarcoplasmic reticulum is made up from the SERCA1 gene product, the RYR1 gene product, and the CAQ1 gene product. In nonmuscle tissues, the endoplasmic reticulum is made up from an alternative isoform from the SERCA2 gene, calreticulin, and the $1P_3$ receptor. Thus each of these systems has unique components. By contrast, slow-twitch muscle is regulated by a hybrid sarcoplasmic reticulum consisting of SERCA2 and phospholamban, RYR1 and CAQ1 gene products.

While our work on cloning of cDNAs encoding sarcoplasmic reticulum proteins has led to a great deal of insight into the structure/function relationships for each of these proteins, in this review we will describe only one aspect of our work relating to our use of cloning, expression, and site-directed mutagenesis to gain insight into the mechanism of Ca^{2+} transport by the Ca^{2+}-ATPase.

THE Ca^{2+}-ATPase

1. Structure

The Ca^{2+}-ATPase is made up of cytoplasmic headpiece and stalk sectors and a transmembrane basepiece, creating a tripartite structure (20). The enzyme is asymmetrically oriented in the membrane with most of its extramembranous mass in the cytoplasm. In our predictions of the topology and the secondary structure of the Ca^{2+}-ATPase that would be consistent with a headpiece, stalk, and basepiece structure, we took into account the fact that NH_2 and COOH-termini, as well as tryptic cleavage and ATP binding sites, are located in the cytoplasm and that disulfide bonds in the protein must exist in the luminal region of the protein. Analysis of hydrophobic sequences led to the assignment of ten trans-membrane helices, four in the NH_2-terminal quarter, and six in the COOH-terminal quarter. These would make up a basepiece with very little protrusion into the luminal space (2, 21). We proposed that the stalk sector would be made up from five alpha helices which are contiguous with transmembrane helices. We also proposed that the headpiece would be made up of three globular domains, the first a seven-membered beta strand domain lying between stalk sectors 2 and 3, and the second and third occurring in alternating alpha-beta sequences lying between stalk sectors 4 and 5. The second and third cytoplasmic domains would be separated by a short linking sequence and rejoined by a long helix at the COOH-terminal end of the third domain. This would form a hinge between the two. We have called these two predicted domains the phosphorylation domain and the nucleotide binding domain.

In our model of the Ca^{2+}-ATPase (Fig. 1) the sequence between residues 758 and 994 makes up six transmembrane helices which form three hairpin loops in the membrane with residues 783–789, 859–896, and 950–962 lying in the lumen (2). This folding model has been supported by the work of Matthews et al. (22) who used monoclonal antibodies to show that the NH_2- and COOH-termini are cytoplasmic. We have used monoclonal antibodies (mAbs) to provide further evidence for this model (23). Antibody A52 (24) reacts with the sequence 657–672 which we predict to be cytoplasmic. We have found that A52 binds

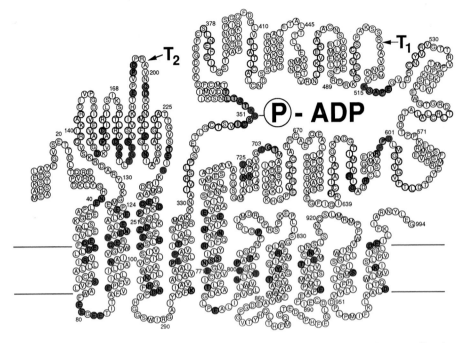

Fig. 1. Structural diagram of the Ca^{2+}-ATPase molecule based on predicted structure and hydropathy plots. The locations of mutated residues are circled and the wild type residues are identified by a single letter code. The functional consequence of each mutation is indicated by a color code: green, no effect on Ca^{2+} transport; yellow, reduced rate of Ca^{2+} transport; red, background level of Ca^{2+} transport; white, no expression of mutant. Note the clustering of red and yellow mutants near the center of the transmembrane domain, where we predict the sites of Ca^{2+} binding to lie, and in the loops between alpha helices and beta strands in the cytoplasmic domain, where we believe the site of ATP binding to lie. Mutations in the stalk sector and in the periphery of the transmembrane domain had little effect on Ca^{2+} transport function.

equally to the ATPase in sarcoplasmic reticulum preparations in the presence and absence of the detergent $C_{12}E_8$, confirming this prediction. A second mAb, A20, reacts with the sequence 870–890. MAb A20 binding to intact sarcoplasmic reticulum vesicles was virtually at background level, but was increased 20 fold in the presence of the detergent, $C_{12}E_8$, or when vesicles were opened by EGTA at elevated pH. These observations demonstrate the localization of the sequence 870–890 on the luminal surface of the sarcoplasmic reticulum and are consistent with 2, 4, or 6 transmembrane sequences in the COOH-terminal quarter. Crystallization of the Ca^{2+}-ATPase is currently being carried out (*25, 26*) and it is probable that analysis of x-ray diffraction patterns will ultimately solve the structure of the protein.

2. *Function*

The Ca^{2+}-ATPase pumps Ca^{2+} against a concentration gradient at the expense of ATP hydrolysis (*27*). During the course of the reaction cycle the enzyme forms a phosphoprotein intermediate and alternates between two conformations referred to as E_1 and E_2. Considerable progress has been made in our understanding of the kinetics of Ca^{2+} binding to and translocation through the Ca^{2+}-ATPase. Two calcium ions bind sequentially, in a highly cooperative fashion, to two high affinity sites on the cytoplasmic face of the sarcoplasmic reticulum membrane (*28*). Ca^{2+} binding is accompanied by conformational changes in the protein. In the E_1 conformation, an acid anhydride bond is found between the γ-phosphate of bound ATP and aspartic acid 351 in the enzyme. The bound Ca^{2+} is distributed into two pools that undergo fast or slow isotopic exchange, respectively (*28, 29*). Both pools of bound Ca^{2+} become inaccessible to quench agents following enzyme phosphorylation by ATP (*30*) and then dissociate sequentially inside the vesicle (*28*). The pool of bound Ca^{2+} which undergoes slower exchange with the outside medium is the first to be released into luminal spaces following enzyme phosphorylation. Translocation and dissociation of bound Ca^{2+} convert the phosphoenzyme from an ADP-sensitive form (E_1P) to an ADP-insensitive form (E_2P). Hydrolysis of the low energy phosphoenzyme, E_2P, and return of the E_2 conformation to the E_1 conformation completes the cycle. The cycle is fully reversible. In the absence of Ca^{2+}, the enzyme can be phosphorylated by inorganic phosphate (*27*) to form an E_2P intermediate. Simultaneous addition of ADP and Ca^{2+} will convert the enzyme to the E_1P form, which will react with ADP to form ATP.

We have used our structural proposals as the basis for an interpretation of the reaction cycle. In our early models we predicted that Ca^{2+} would bind to the stalk domain, since the first three stalk helices are markedly amphipathic, each with 4–6 aligned glutamic acid residues on the polar face (*2, 31*). ATP, bound on the surface of the headpiece, would then phosphorylate Asp351. Phosphorylation-induced rotation of parts of the stalk sector might occlude the Ca^{2+} and the accompanying disruption of the high affinity sites might allow the Ca^{2+} to escape (*32*), but only into a channel formed by stalk and transmembrane sequences leading to the other side of the membrane.

We felt that one way to improve our understanding of the molecular details of the Ca^{2+} transport mechanism would be to identify, through site-directed mutagenesis, those residues involved in the specific partial reactions of the pump cycle. This approach has been facilitated by the development of an efficient system for mutagenesis and expression of the Ca^{2+}-ATPase gene product (*33*).

3. *Expression and Mutagenesis*

We used three different approaches in our early attempts to express the full-length Ca^{2+}-ATPase cDNA in a functional form: (1) stable transfection in mammalian cells; (2) transient transfection in mammalian cells; and (3) expression in prokaryotic cells. Expression of the Ca^{2+}-ATPase cDNA was observed in all three systems. The expressed Ca^{2+}-ATPase was rapidly degraded in *Escherichia coli* cells, however, making the system unsuitable for our purposes. By contrast, the expressed Ca^{2+}-ATPase is not subject to rapid proteolytic breakdown when expressed in mammalian cell lines. Transient expression in COS-1 cells has been the system of choice, since this method allows more rapid assay of mutations and gives higher expression levels than stable transfection. In our expression system we have inserted the full-length cDNA into the *Eco*R1 site of the vector p91023(B) (*34*). Because COS-1 cells express the T antigen (*35*), and because the vector has an SV40 origin of replication, it is replicated in high copy number in transfected cells. Microsomes prepared from cells transfected with Ca^{2+}-ATPase cDNA can exhibit Ca^{2+} transport activities more than 50-fold higher than background. In addition, we can monitor partial reactions of the expressed Ca^{2+}-ATPase, such as formation of the E_1P or E_2P intermediates and the E_1P to E_2P conformational transitions, by incubating the microsomes containing the expressed Ca^{2+}-ATPase with radioactive ATP or inorganic phosphate under the appropriate conditions. Our ability to monitor partial reactions of the pump cycle in expressed Ca^{2+}-ATPases has greatly facilitated our analysis of mutants inactive in overall Ca^{2+} transport. In Fig. 1 we present a summary of our mutations in the Ca^{2+}-ATPase on the background of our predicted model for its structure.

4. *Analysis of the Phosphorylation and ATP Binding Domains*

Location of the catalytic site in the cytosolic portion of the ATPase was established by the identification of Asp351 as the residue undergoing phosphorylation upon ATP utilization (*36, 37*). In our first study (*33,*

38), we investigated the effects of mutations in the sequence ICSDKTGT-LT357, which contains the catalytic site Asp351 and is highly conserved in all P-type cation pumps. Changes at the catalytic site abolished both formation of the phosphoenzyme intermediate and Ca^{2+} transport, a result which confirms the central role of this residue in the reaction scheme. All mutations of Lys352 also disrupted Ca^{2+} transport function and phosphoenzyme formation. Alterations to the residues surrounding Asp351 and Lys352 indicated that highly conservative replacements could be tolerated without disrupting function. Nonconservative substitutions for most of these residues, however, disrupted both transport and phosphorylation in parallel. These results suggest that the highly conserved residues surrounding Asp351 play critical roles in phosphoryl transfer and hydrolytic reaction steps of the ATPase cycle.

Another region with a high degree of amino acid sequence identity among the family of cation pumps occurs at the COOH-terminal end of the proposed cytoplasmic domain. The sequence has been implicated by chemical labeling in the formation of part of the nucleotide binding domain. In models presented by Taylor and Green (*39*) and Serrano (*40*), Lys515, Asp601, Asp627, Asp703, and Asp707 are grouped around a postulated ATP-binding cleft. We have constructed mutants with alterations to these and surrounding amino acids (*33, 38, 41*). Mutants Asp601 →Glu, Pro603→Gly, Gly626→Ala, Gly626→Pro, Asp707→Asn, and Asp707→Glu did not form detectable phosphoenzyme intermediates in the presence of either ATP or P_i. These results are consistent with, but do not prove the involvement of these residues in ATP binding. All other substitutions did form phosphoenzyme intermediates in the presence of ATP, indicating that the ATP was bound in the molecule. Our results suggest that the conserved residues in the proposed nucleotide binding domain play important roles in the structure of the phosphorylation and nucleotide binding domains and in conformational changes occurring after phosphorylation with ATP.

5. *Mutational Analysis of Ca^{2+} Binding Sites*

In contrast to the catalytic site, the probable location of the calcium-binding domain was unknown when we began our mutagenesis studies. The amino acid sequence of the Ca^{2+}-ATPase does not include stretches homologous to consensus binding sequences found in other calcium binding proteins. A candidate region, however, was the cluster of acidic residues that we found in the amphipathic helices of the stalk sector which connect the cytoplasmic and intramembranous portions of the ATPase. Accordingly, we mutated all of the Glu, Gln, Asp, and Asn residues in this sector (*42*). Of more than twenty-five residues altered either individually or in groups, few had any significant influence on Ca^{2+} transport. Measurement of the Ca^{2+}-dependency of Ca^{2+} transport, an approximate measurement of Ca^{2+} binding affinity, confirmed that none of these mutations altered Ca^{2+} affinity. One mutant Asn111→Ala, exhibited only 10% of wild-type Ca^{2+} transport function. This mutant was phosphorylated by ATP in the presence of Ca^{2+} as effectively as wild type and no phosphorylation was observed in the absence of Ca^{2+}. Thus the Ca^{2+} binding sites of this mutant remained intact. We concluded that none of the acidic or amidated residues in stalk or luminal sectors 1, 2, 3, or 5 is involved in a critical way in Ca^{2+} binding.

We then turned our attention to the acidic residues of the transmembrane domain which could participate as ligands at the high affinity Ca^{2+} binding sites. Changes to four acidic residues, Glu309, Glu771, Asp800, or Glu908 abolished Ca^{2+} transport function and the formation of the phosphoenzyme intermediate from ATP (*43*). On the other hand, the phosphorylation domain remained intact, since the mutant enzymes could be phosphorylated with P_i, a reaction which occurs only when the Ca^{2+} binding sites are unoccupied. In fact, addition of Ca^{2+} to the mutated ATPases did not inhibit phosphorylation by P_i. Thus, abolition of phosphorylation of the enzyme from ATP in the presence of Ca^{2+} and/or prevention of phosphorylation from P_i in the presence of Ca^{2+} were consistent with the loss of high affinity Ca^{2+} binding sites in these mutants. Mutations have now been made to most of the residues in transmembrane sequences which contain oxygen groups in their side chains and which could provide ligands for high affinity Ca^{2+} binding (*43, 44*). Of these, Asn796 and Ser799 were also identified as potential Ca^{2+} binding ligands since mutations to these residues resulted in mutants with properties similar to those observed for mutants with changes to Glu309, Glu771, Asp800, or Glu908. Alterations to Ser766, Ser767, and Asn768 did not abolish activity but influenced the apparent affinity of the enzyme for Ca^{2+}. This suggests either that these residues provide some of the ligands for the Ca^{2+} binding sites, or that they are important in maintaining the correct conformation of the binding ligands. Mutation of two proline residues (Pro308 and Pro803) also modulates the affinity of the enzyme for Ca^{2+} (*45*). This is likely a consequence of the requirement for maintenance of the alignment of other residues close by those which form the Ca^{2+} binding site.

Evidence supporting a relationship between acidic residues residing within the transmembrane domain and specific calcium binding has recently been provided by chemical labeling studies. Inesi *et al.* (*46*) demonstrated that labeling of Ca^{2+}-ATPase with N-cyclohexyl-N-(4-di-methylaminonaphthyl)carbodiimide (NCD_4) inhibits Ca^{2+} binding. The loss of Ca^{2+} binding after enzyme derivatization with NCD_4 is accompanied by loss of the two specific functional effects of Ca^{2+}, phosphorylation by ATP, and inhibition of phosphorylation with P_i. Fluorescence energy transfer experiments indicate that the label lies within or near the transmembrane domain of the ATPase.

On the basis of these observations, we propose that the high affinity Ca^{2+} binding sites are located near the center of the transmembrane domain and that the carboxylate or carboxamide side chains of Glu309, Glu771, Asn768, Asn796, Asp800, and Glu908 and the hydroxyl groups of Ser766, Ser767, and Thr799 act as ligands to form the high affinity Ca^{2+} binding sites.

6. Model for Ca^{2+} Transport

The data presented in Fig. 1 illustrate an interesting point about our overall studies of site-directed mutagenesis. Of those mutations in the transmembrane sector, only a small group clustered around the centers of transmembrane sequences 4, 5, 6, and 8 affect Ca^{2+} binding. In the color coding that we have used (red means loss of transport; yellow means reduced transport rates; green means no effect on transport) the red and yellow cluster is surrounded by a large number of green residues where mutations had no effect on Ca^{2+} transport. In analogy with city zoning, we can imagine that this warmly colored cluster represents a small "industrial zone," dedicated to the binding and transport of Ca^{2+}, while the surrounding "green belt" provides a different supporting structure. If we focus on the headpiece domain, we again see an industrial park where ATP is bound and hydrolyzed and this too is surrounded by a green belt of residues less directly involved at the catalytic site. Nevertheless, the two industrial parks are in communication through a syncytium of residues which are critical to the conformational changes that carry signals between the two centers, informing each other of the occupancy of the respective binding sites by ATP and Ca^{2+}. We have identified some of these residues, since mutants in any of them blocks the conformational change between E_1P and E_2P. We refer to these residues, found throughout the three cytoplasmic domains, the stalk sector and the basepiece, as conformational change mutants (44, 45, 47, 48).

These concepts concerning Ca^{2+} and ATP binding site and the conformational changes linking the two have allowed us to develop an easily understood model for Ca^{2+} transport which is illustrated in Fig. 2. When the Ca^{2+}-ATPase is in the E_1 conformation, cytoplasmic Ca^{2+} has diffusional access to two high affinity binding sites near the center of the transmembrane domain. These binding sites act as a gate, preventing entry of cytoplasmic Ca^{2+} to the lumen or exit of luminal Ca^{2+} to the cytoplasm. Occupancy of one Ca^{2+} binding site, formed by oxygen ligands from amino acids in different transmembrane segments, results in a reorientation of the segments, which increases the affinity of the second site for Ca^{2+}. The presence of Ca^{2+} in both binding sites leads to conformational changes which are transmitted to the nucleotide and phosphorylation domains to bring the bound ATP and Asp351 into a reactive configuration. After the transfer of phosphate, further confor-

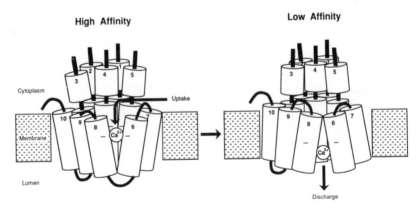

Fig. 2.   Model illustrating the mechanism of Ca^{2+} transport by the Ca^{2+}-ATPase. In the E_1 conformation, high affinity Ca^{2+} binding sites located near the center of the transmembrane domain are accessible to cytoplasmic Ca^{2+}, but not to luminal Ca^{2+} (high affinity state). The sites are made up from amino acid residues located in proposed transmembrane sequences M4, M5, M6, and M8. Conformational changes induced by ATP hydrolysis lead to the E_2 conformation in which the high affinity Ca^{2+} binding sites are disrupted, access to the sites by cytoplasmic Ca^{2+} is closed off and access to the sites by luminal Ca^{2+} is gained (low affinity state). The Ca^{2+} transport cycle thus involves binding of cytoplasmic Ca^{2+} to high affinity sites in one conformation and release of the same Ca^{2+} to the lumen when the high affinity sites are disrupted in the transition to the second conformation.

mational motion leads to rotation or tilting of one or more of the trans-membrane segments, disrupting the high affinity Ca^{2+} binding domains and changing the orientation of the Ca^{2+} pair so that they become inaccessible from the external medium, but accessible to a channel leading to the luminal side of the membrane. The diffusional movement of Ca^{2+} from the disrupted binding sites to the lumen completes the cycle of Ca^{2+} transport.

Coupling between the phosphorylation and calcium binding domains probably involves a large portion of the protein, since coupling is trans-mitted over a large distance. The actual conformational changes are probably small, however, since it is known that large changes in second-ary structure do not occur in the Ca^{2+}-ATPase (*49*). The nature of these conformational changes remains an important goal for our further understanding of Ca^{2+} transport.

Acknowledgments

Original research described in this review was supported by grants to DHM from the Medical Research Council of Canada, the National Institutes of Health (U.S.A.), the Muscular Dystrophy Association of Canada, and the Heart and Stroke Foundation of Ontario. The technical assistance of Kazimierz Kurzydlowski in this work is gratefully acknowledged.

REFERENCES

1. MacLennan, D.H., Brandl, C.J., Champaneria, S., Holland, P.C., Powers, V.E., and Willard, H.F. *Som. Cell Mol. Genet.*, **13**, 341 (1987).
2. Brandl, C.J., Green, N.M., Korczak, B., and MacLennan, D.H. *Cell*, **44**, 597 (1986).
3. Lytton, J. and MacLennan, D.H. *J. Biol. Chem.*, **263**, 15024 (1988).
4. Lytton, J., Zarain-Herzberg, A., Periasamy, M., and MacLennan, D.H. *J. Biol. Chem.*, **264**, 7059 (1989).
5. Jorgensen, A.O. and Jones, L.R. *J. Cell Biol.*, **104**, 1343 (1987).
6. Fujii, J., Zarain-Herzberg, A., Willard, H.F., Tada, M., and MacLennan, D.H. *J. Biol. Chem.*, **266**, 11669 (1991).
7. Fliegel, L., Ohnishi, M., Carpenter, M.R., Khanna, V.K., Reithmeier, R.A.F., and MacLennan, D.H. *Proc. Natl. Acad. Sci. U.S.A.*, **84**, 1167 (1987).
8. Fliegel, L., Leberer, E., Green, N.M., and MacLennan, D.H. *FEBS Lett.*, **242**, 297 (1989).
9. Fujii, J., Willard, H.F., and MacLennan, D.H. *Som. Cell Mol. Genet.*, **16**, 185 (1990).
10. Leberer, E., Charuk, J.H.M., Clarke, D.M., Green, N.M., Zubrzycka-Gaarn, E., and MacLennan, D.H. *J. Biol. Chem.*, **264**, 3484 (1989).
11. Leberer, E., Charuk, J.H.M., Green, N.M., and MacLennan, D.H. *Proc. Natl. Acad. Sci. U.S.A.*, **86**, 6047 (1989).

12. Hofmann, S.L., Brown, M.S., Lee, E., Pathak, R.K., Anderson, R.G.W., and Goldstein, J.L. *J. Biol. Chem.*, **264**, 8260 (1989).
13. Hofmann, S.L., Goldstein, J.L., Orth, K., Moomaw, C.R., Slaughter, C.A., and Brown, M.S. *J. Biol. Chem.*, **264**, 18083 (1989).
14. Scott, B., Simmerman, H.K.B., Collins, J.H., Nadal-Ginard, B., and Jones, L.R. *J. Biol. Chem.*, **263**, 8459 (1988).
15. Fliegel, L., Burns, K., MacLennan, D.H., Reithmeier, R.A.F., and Michalak, M. *J. Biol. Chem.*, **264**, 21522 (1989).
16. McCouliffe, D.P., Lux, F.A., Lieu, T.-S., Sanz, I., Hanke, J., Newkirk, M.M., Bachinski, L.L., Itoh, Y., Siciliano, J., Reichlin, M., Sontheimer, R.D., and Capra, J.D. *J. Clin. Invest.*, **85**, 1379 (1990).
17. MacKenzie, A.E., Korneluk, R.G., Zorzato, F., Fujii, J., Phillips, M., Iles, D., Wieringa, B., Leblond, S., Bailly, J., Willard, H.F., Duff, C., Warton, R.G., and MacLennan, D.H. *Am. J. Hum. Genet.*, **46**, 1082 (1990).
18. Otsu, K., Willard, H.F., Khanna, V.J., Zorzato, F., Green, N.M., and MacLennan, D.H. *J. Biol. Chem.*, **265**, 13472 (1990).
19. Furuichi, T., Yoshikawa, S., Miyawaki, A., Wada, K., Maeda, N., and Mikoshiba, K. *Nature*, **342**, 32 (1989).
20. MacLennan, D.H. and Reithmeier, R.A.F. *In* "Structure and Function of Sarcoplasmic Reticulum," ed. S. Fleischer and Y. Tonomura, p. 91 (1985). Academic Press, New York.
21. MacLennan, D.H., Brandl, C.J., Korczak, B., and Green, N.M. *Nature*, **316**, 696 (1985).
22. Matthews, I., Colyer, I., Mata, A.M., Green, N.M., Sharma, R.P., Lee, A.G., and East, J.M. *Biochem. Biophys. Res. Commun.*, **161**, 683 (1989).
23. Clarke, D.M., Loo, T.W., and MacLennan, D.H. *J. Biol. Chem.*, **265**, 17405 (1990).
24. Zubrzycka-Gaarn, E., MacDonald, G., Phillips, L., Jorgensen, A.O., and MacLennan, D.H. *J. Bioenergetics Biomembranes*, **16**, 441 (1984).
25. Stokes, D.L. and Green, N.M. *Biophys. J.*, **57**, 1 (1990).
26. Taylor, K.A., Mullner, N., Pikula, S., Dux, L., Peracchia, C., Varga, S., and Martonosi, A. *J. Biol. Chem.*, **263**, 5287 (1988).
27. deMeis, L. and Vianna, A. *Annu. Rev. Biochem.*, **48**, 275 (1979).
28. Inesi, G. *J. Biol. Chem.*, **262**, 16338 (1987).
29. Nakamura, J. *J. Biol. Chem.*, **262**, 14492 (1987).
30. Takisawa, H. and Makinose, M. *J. Biol. Chem.*, **258**, 2986 (1983).
31. MacLennan, D.H., Brandl, C.J., Korczak, B., and Green, N.M. *In* "Proteins of Excitable Membranes," ed. B. Hille and D. Fambrough, p. 287 (1986). Wiley, New York.
32. Tanford, C. *Proc. Natl. Acad. Sci. U.S.A.*, **79**, 2882 (1982).
33. Maruyama, K. and MacLennan, D.H. *Proc. Natl. Acad. Sci. U.S.A.*, **85**, 3314 (1988).
34. Wong, G.G., Witek, J.S., Temple, P.-A., Wilkens, K.M., Leary, A.C., Luxenberg, D.P., Jones, S.S., Brown, E.L., Kay, R.M., Orr, E.C., Shoemaker, C., Golde, D.W., Kaufman, R.J., Hewick, R.M., Wang, E.A., and Clark, S.C. *Science*, **28**, 810 (1985).
35. Gluzman, Y. *Cell*, **23**, 175 (1981).
36. Bastide, F., Meissner, G., Fleischer, S., and Post, R.L. *J. Biol. Chem.*, **248**, 8385 (1973).
37. Degani, C. and Boyer, P.D. *J. Biol. Chem.*, **248**, 8222 (1973).
38. Maruyama, K., Clarke, D.M., Fujii, J., Inesi, G., Loo, T.W., and MacLennan, D.H. *J. Biol. Chem.*, **264**, 13038 (1989).

39. Taylor, W.R. and Green, N.M. *Eur. J. Biochem.*, **179**, 241 (1989).
40. Serrano, R. *Annu. Rev. Plant Physiol. Mol. Biol.*, **40**, 61 (1989).
41. Clarke, D.M., Loo, T.W., and MacLennan, D.H. *J. Biol. Chem.*, **265**, 22233 (1990).
42. Clarke, D.M., Maruyama, K., Loo, T.W., Leberer, E., Inesi, G., and MacLennan, D.H. *J. Biol. Chem.* **264**, 11246 (1989).
43. Clarke, D.M., Loo, T.W., Inesi, G., and MacLennan, D.H. *Nature*, **339**, 476 (1989).
44. Clarke, D.M., Loo, T.W., and MacLennan, D.H. *J. Biol. Chem.*, **265**, 6262 (1990).
45. Vilsen, B., Andersen, J.P., Clarke, D.M., and MacLennan, D.H. *J. Biol. Chem.*, **264**, 21024 (1989).
46. Inesi, G., Sumbilla, C., and Kirtley, M.E. *Physiol. Rev.*, **70**, 749 (1990).
47. Andersen, J.P., Vilsen, B., Leberer, E., and MacLennan, D.H. *J. Biol. Chem.*, **264**, 21018 (1989).
48. Clarke, D.M., Loo, T.W., and MacLennan, D.H. *J. Biol. Chem.* **265**, 14088 (1990).
49. Nakamoto, R.K. and Inesi, G. *FEBS Lett.*, **194**, 258 (1986).

Possible Functions and Mode of Action of *smg* p21 in Signal Transduction

YOSHIMI TAKAI,*¹ KOZO KAIBUCHI,*¹ AKIRA KIKUCHI,*¹
MASAHITO KAWATA,*¹ TAKUYA SASAKI,*¹ AND
YASUHIRO KAWAHARA*²

*Departments of Biochemistry*¹ *and Internal Medicine (1st Division),*² *Kobe University School of Medicine, Kobe 650, Japan*

There is a superfamily of *ras* p21/*ras* p21-like small GTP-binding proteins (G proteins) in addition to a superfamily of G proteins serving as transducers for membrane receptors such as G_s, G_I, G_o, and transducin (*1*, *2*). The members of small G proteins of which primary structures have been reported in the literature are listed in Table I. All of these small G proteins have the consensus amino acid sequences for GDP/GTP-binding and GTPase activities; several of them have been purified from mammalian tissues or synthesized in *Escherichia coli*. The purified proteins indeed exhibit GDP/GTP-binding and GTPase activities. Among these small G proteins, *ras* p21 has been most extensively investigated (*1*), and has been shown to exhibit cell transforming and differentiating activities depending on the cell type (*1*). However, the mode of action of *ras* p21 remains to be clarified. The *SEC*4 and *YPT*1 proteins are involved in the secretory processes in yeast (*3*, *4*), but in this case again the mode of action is not known. Neither the functions nor the modes of action of other small G proteins are understood.

In our laboratory, we have separated many small G proteins from mammalian tissues and discovered two novel small G protein families referred to as *smg* p21 and *smg* p25 families (*5–8*). The *smg* p21 family is particularly abundant in aortic smooth muscle, heart, and platelets (*9*, *10*), and is composed of two highly homologous members, A and B. Both *smg* p21A and -B are composed of 184 amino acids with calculated M_r values of about 21,000 and differ by only 9 amino acids (Fig. 1). *smg*

15

TABLE I
Small G Proteins

Gene	M_r value
Ha-*ras*	21K
Ki-*ras*	21K
N-*ras*	21K
*rho*A	21K
*rho*B	21K
*rho*C	21K
ral	23K
R-*ras*	23K
*SEC*4	24K
*ypt*1 = *rab*1	24K
*rab*2	24K
*rab*4	24K
*rab*5	23K
*rab*6	24K
*arf*1	21K
*arf*2	21K
*arf*3	21K
*rac*1	21K
*rac*2	21K
smg-25A = *rab*3A	25K
smg-25B = *rab*3B	25K
smg-25C	26K
smg-21A = *rap*1A = K*rev*-1	21K
smg-21B = *rap*1B	21K

p21A is identical to the *rap*1A and K*rev*-1 proteins and *smg* p21B is identical to the *rap*1B protein (*7*, *8*, *11–13*). In this review article, the possible functions and modes of action and activation of *smg* p21 are described.

I. TISSUE AND SUBCELLULAR DISTRIBUTIONS OF *smg* p21

We have made a polyclonal antibody specifically recognizing both *smg* p21A and -B (*14*) with which we have compared the tissue and subcellular distributions of *smg* p21 with those of *ras* p21. The two families are ubiquitous in most mammalian tissues but their tissue and subcellular distributions differ somewhat (*14*). For instance, *smg* p21 is abundant in rat testis while little *ras* p21 is detected. In bovine aortic smooth muscle, rat heart, and human platelets, the amount of *smg* p21 is several tens of times greater than that of *ras* p21 (*9*, *10*). In rat brain, *smg* p21 is found in both the synaptic and neuron body areas, while *ras* p21

smg p21A

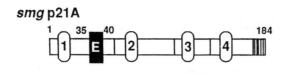

smg p21B

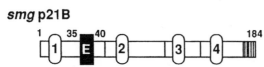

Fig. 1.   Primary structures of *smg* p21A and -B. 1 and 2, GTPase domain; 3, and 4, guanine nucleotide-binding domain; E, effector domain. Lines indicate the different amino acids between *smg* p21A and -B.

is found mainly in the synaptic area (*14*, *15*). In the synapses, *smg* p21 is found in the synaptic plasma membrane, the synaptic vesicle, and the synaptic mitochondria, while *ras* p21 is primarily located in the synaptic plasma membrane (*14*, *15*).

II.   MODE OF ACTIVATION OF *smg* p21

It is conceivable by analogy with *ras* p21 that *smg* p21 has two convertible GDP-bound inactive and GTP-bound active forms (Fig. 2). We have recently purified from bovine brain cytosol a GDP/GTP exchange protein for *smg* p21A and -B that stimulates the dissociation of GDP from and the subsequent binding of GTP to the proteins without affecting its GTPase activity (*16*) (Fig. 3A and B). This regulatory protein, called GDP dissociation stimulator (GDS), is active on *smg* p21A and -B and is inactive on other small G proteins including c-Ha-*ras* p21 and *smg* p25A. *smg* GDS is found in most mammalian tissues and is mainly located in the cytosol (*17*). The M_r value of *smg* GDS is about 53,000 as estimated by sodium dodecylsulfate polyacrylamide gel electrophoresis and the S value. Another type of GDP/GTP exchange protein for *smg* p25A and *rho*p21, named GDP dissociation inhibitor (GDI), was also discovered (*18-20*). This type of regulatory protein inhibits the dissociation of GDP from and thereby the subsequent binding of GTP to each G protein. *smg* p25 GDI is active on *smg* p25A and 24KG purified from human platelets and rat liver, both of which are similar to but not identical to *smg* p25A (*21*, *22*). *rho* GDI is active on both *rho*A p21 and *rho*B p21 (*19*, *20*). *smg* GDS has no GDP/GTP-binding, GTPase, or GDI activity.

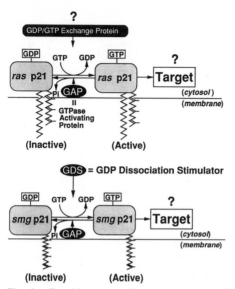

Fig. 2. Possible modes of activation and action of *ras* p21 and *smg* p21.

In addition to this *smg* GDS, we have partially purified a regulatory protein for *smg* p21 that stimulates the GTPase activity without affecting

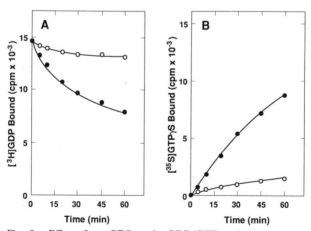

Fig. 3.   Effect of *smg* GDS on the GDP/GTP exchange reaction of *smg* p21B. The activities of *smg* GDS to stimulate the dissociation of [³H]GDP from *smg* p21B and the binding of [³⁵S]GTPγS to the GDP-bound form of *smg* p21B were measured in the presence or absence of *smg* GDS. Detailed conditions are described (*16*). A, the dissociation of [³H]GDP; B, the binding of [³⁵S]GTPγS. ● in the presence of *smg* GDS; ○ in the absence of *smg* GDS.

the GDP/GTP exchange reaction (*23, 24*). This *smg* p21 GAP is specific for this small G protein and is found in the cytosol of most mammalian tissues. Thus, the activation and inactivation of *smg* p21 are regulated by its specific GDS and GAP, respectively.

III. THE SAME EFFECTOR DOMAIN OF *smg* p21 AS THAT OF *ras* p21

Among many small G proteins, only *smg* p21A and -B have the same putative effector domain as that of *ras* p21 (Fig. 1). This structural property theoretically suggests that the two families can share the same effector proteins. Consistently, the K*rev*-1 gene has been shown to suppress the transforming activity of the activated Ki-*ras* gene in NIH/3T3 cells (*13*). We have also found that the *smg* p21B purified from human platelets interacts with *ras* p21 GAP and inhibits its activity (*25*) (Fig. 4). *ras* p21 GAP interacts with at least the effector domain of *ras* p21 (*26*). Therefore, it is likely that *smg* p21 competes with *ras* p21 for the interaction with *ras* p21 GAP and inhibits its activity. Since *ras* p21 GAP is

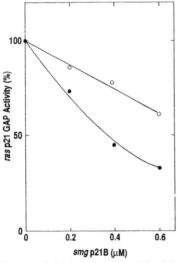

Fig. 4.   Inhibition by *smg* p21B of the *ras* p21 GAP-stimulated GTPase activity of c-Ha-*ras* p21. The activity of *ras* p21 GAP to stimulate the liberation of $^{32}P_i$ from the [γ-^{32}P]GTP-bound form of c-Ha-*ras* p21 was measured in the presence of various amounts of the GTP- or GDP-bound form of *smg* p21B. Detailed conditions are described (*25*). Activity is expressed as a percent of the activity of *ras* p21 GAP measured in the absence of *smg* p21B. ● the GTP-bound form; ○ the GDP-bound form.

inactive on *smg* p21 and *smg* p21 GAP is inactive on *ras* p21, each GAP may interact with other regions in addition to the effector domain of each G protein.

IV. A UNIQUE C-TERMINAL STRUCTURE OF *smg* p21

All small G proteins have unique C-terminal structures and are classified into three groups according to their C-terminal sequences (*2*) (Fig. 5). *ras* p21, *smg* p21, and several other small G proteins have a Cys-A-A-X structure where A is an aliphatic amino acid and X is any amino acid. *ras* p21 has been shown to be posttranslationally processed at this C-terminal region (*27*) (Fig. 6). Firstly, the cysteine residue in the C-terminal region is farnesylated. This farnesyl group is derived from farnesylpyrophosphate by the action of farnesyltransferase. Farnesyl-pyrophosphate is an intermediate product of the cholesterol synthesis from mevalonate. Then, the three C-terminal amino acids are removed by proteolysis and the exposed C-terminal cysteine residue is carboxyl methylated. Finally, in the case of c-Ha- and N-*ras* p21s, the cysteine residue near the C-terminal cysteine residue is palmitoylated (Fig. 6). In the case of c-Ki-*ras* p21, this second cysteine residue is not present and is not palmitoylated. The farnesylation is essential for *ras* p21 to bind to membranes and to exert its biological action (*27*). In contrast to *ras* p21, we have found in collaboration with Drs. J.A. Glomset and M.H. Gelb

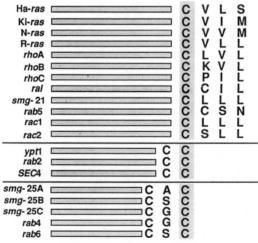

Fig. 5.   Three types of C-terminal sequences of small G proteins.

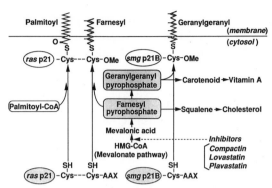

Fig. 6.   Posttranslational modifications of *ras* p21 and *smg* p21B.

(University of Washington, Seattle, U.S.A.) that the *smg* p21B purified from human platelets is geranylgeranylated at its C-terminal cysteine residue, that the C-terminal three amino acids are removed, and that the exposed cysteine residue is carboxyl methylated (*28*) (Fig. 6). This lipid modification is essential for *smg* p21B to bind to membranes (*29*).

V.   PHOSPHORYLATION OF *smg* p21 BY PROTEIN KINASE A

It has previously been reported that *ras* p21 is phosphorylated by protein kinases A and C in both cell-free and intact cell systems (*30, 31*), but the stoichiometry of P_i incorporated into *ras* p21 has not been completely defined. In our analysis, the c-Ki-*ras* p21 purified from bovine brain membranes is not phosphorylated by protein kinase C at all and is slightly phosphorylated by protein kinase A as far as tested in a cell-free system (*32*). Less than 0.1 mol/mol of protein is phosphorylated, so that the physiological significance of this action of *ras* p21 remains to be clarified.

On the other hand, we have found that the *smg* p21B purified from bovine brain membranes and human platelet membranes is phosphorylated by protein kinase A but not by other protein kinases including protein kinase C, calmodulin-dependent multifunctional protein kinase, the CDC2 kinase, the platelet-derived growth factor (PDGF) receptor tyrosine kinase, the epidermal growth factor (EGF) receptor tyrosine kinase, or the insulin receptor tyrosine kinase (*32, 33*). Among many small G proteins in bovine brain membranes, only *smg* p21B is phosphorylated (*32*). This protein phosphorylation, indeed, occurs in intact platelets in response to dibutyryl cyclic AMP or cyclic AMP-elevating prostaglandin E_1 (*32*). We have recently found that the *smg* p21A purified from bovine aortic smooth muscle is also phosphorylated by protein kinase A (*34*).

The site of the phosphorylation of *smg* p21B is Ser^{179}, and is near the C-terminal cysteine residue (*35*). The phosphorylation of *smg* p21B does not affect its GDP/GTP-binding or GTPase activity, nor does it affect the sensitivity of *smg* p21B to the *smg* p21 GAP action, but makes *smg* p21B sensitive to the GDS action. Namely, when *smg* p21B is phosphorylated by protein kinase A, *smg* GDS greatly enhances the GDP/GTP exchange reaction (*35*) (Fig. 7). These results suggest that the cyclic AMP-protein kinase A system induces the activation of *smg* p21. Moreover, the C-terminal region of *smg* p21 appears to be important not only for its binding to membranes but also for its regulation by *smg* GDS.

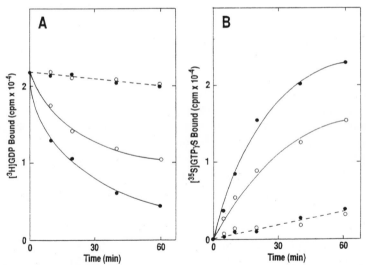

Fig. 7. *smg* GDS activities for the phosphorylated and unphosphorylated forms of *smg* p21B. *smg* p21B was phosphorylated by protein kinase A as described (*32, 33*). The activities of *smg* GDS for the phosphorylated or unphosphorylated form of *smg* p21B were assayed in the presence or absence of *smg* GDS as described in the legend to Fig. 3. Detailed conditions are described (*35*). A, the dissociation of [³H]GDP; B, the binding of [³⁵S]GTPγS. ● the phosphorylated form; ○ the unphosphorylated form. —— in the presence of *smg* GDS; ---- in the absence of *smg* GDS.

VI. POSSIBLE FUNCTIONS OF *smg* p21

The results of the tissue and subcellular distributions of *smg* p21 and *ras* p21 and the fact that *smg* p21 has the same effector domain as that of *ras* p21 described above suggest that *smg* p21 exerts actions similar or antagonistic to those of *ras* p21 in places where both are present, and that *smg* p21 exerts its specific actions in places where it alone is present. In fact, the K*rev*-1 gene has been shown to suppress the Ki-*ras* p21 transforming activity in NIH/3T3 cells as described above (*13*). We and others have also shown that *smg* p21B and the *rap*1A protein inhibit the *ras* p21 GAP activity as described above (*25, 36*). It is most likely that *ras* p21 GAP serves as a regulatory protein for *ras* p21 that converts the GTP-bound active form to the GDP-bound inactive form. If this is the case, *smg* p21 may inhibit the *ras* p21 GAP activity, keeping *ras* p21 in the active form, and enhancing the *ras* p21 actions. However, the possibility still exists that *ras* p21 GAP may serve as an effector protein since *ras* p21 GAP interacts with the effector domain of *ras* p21 (*26*). If this is the case, *smg* p21 may antagonize the *ras* p21 action and this mode of action would be consistent with the results obtained with the K*rev*-1 gene (*13*). We assume that these apparently opposite actions of *smg* p21 may be possible depending on cell types and its intracellular localization.

It has recently been reported that *ras* p21 GAP is phosphorylated at its tyrosine residue by the PDGF and EGF receptor tyrosine kinases (*37–39*). It has also been shown that *ras* p21 GAP interacts with these receptors which are autophosphorylated at their tyrosine residue (*37–39*). *ras* p21 GAP has been shown to have a *src* homology region 2 (SH2) domain and this domain may recognize the phosphotyrosine of the PDGF receptor and bind to it. Although the physiological significance of this phosphorylation is not known, it is likely that the actions of the PDGF receptors may be at least partly mediated by *ras* p21. These observations together with our results suggest that *smg* p21 may play an important role in the protein tyrosine kinase system (Fig. 8).

We have also shown that *smg* p21 is phosphorylated by protein kinase A as described above. In a certain type of cells, protein kinase C activation and Ca^{2+} mobilization induced by the receptor-linked activation of the phosphoinositide-specific phospholipase C cause the cellular activation and this is antagonized by the cyclic AMP-protein kinase A system (*40*). For instance, in platelets the secretion is induced by the

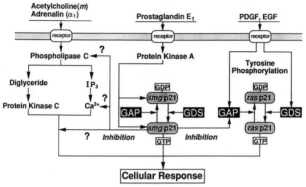

Fig. 8.   *smg* p21 and intracellular messenger systems.

cooperative role of protein kinase C and Ca²⁺ and this release reaction is inhibited by cyclic AMP. In smooth muscle, protein kinase C and Ca²⁺ induce contraction and cyclic AMP induces relaxation. Since *smg* p21 is phosphorylated by protein kinase A, it is conceivable that it may play a role in the signal transduction system of the cyclic AMP systems , and also in crosstalk between this system and the protein kinase and Ca²⁺ system. Thus, evidence is accumulating that *smg* p21 plays important roles in positive and/or negative cooperation with well-known intracellular messenger systems (Fig. 8).

SUMMARY

There is a superfamily of small G proteins in mammalian tissues among which *smg* p21 is particularly abundant in vascular smooth muscle, heart, and platelets. It is likely that *smg* p21 plays important roles in signal transduction in positive and/or negative cooperation with well-known intracellular messenger systems such as the protein tyrosine kinase, protein kinase C, Ca²⁺, and cyclic AMP-systems. Further investigation is necessary to establish the physiological functions and modes of activation and action of *smg* p21 and other small G proteins in signal transduction.

REFERENCES

1. Barbacid, M. *Annu. Rev. Biochem.*, **56**, 779 (1987).
2. Takai, Y., Kikuchi, A., Yamashita, T., Yamamoto, K., Kawata, M., and Hoshijima, M.

In "Progress in Endocrinology," Vol. 2, ed. H. Imura, K. Shizume, and S. Yoshida, p. 995 (1988). Elsevier Science Publishers, B.V., Amsterdam.

3. Gove, B., Salminen, A., Walworth, N.C., and Novick, P.J. *Cell*, **53**, 753 (1988).
4. Segev, N., Mulholland, J., and Botstein, D. *Cell*, **52**, 915 (1988).
5. Kikuchi, A., Yamashita, T., Kawata, M., Yamamoto, K., Ikeda, K., Tanimoto, T., and Takai, Y. *J. Biol. Chem.*, **263**, 2897 (1988).
6. Matsui, Y., Kikuchi, A., Kondo, J., Hishida, T., Teranishi, Y., and Takai, Y. *J. Biol. Chem.*, **263**, 11071 (1988).
7. Kawata, M., Matsui, Y., Kondo, J., Hishida, T., Teranishi, Y., and Takai, Y. *J. Biol. Chem.*, **263**, 18965 (1988).
8. Matsui, Y., Kikuchi, A., Kawata, M., Kondo, J., Teranishi, Y., and Takai, Y. *Biochem. Biophys. Res. Commun.*, **166**, 1010 (1990).
9. Ohmori, T., Kikuchi, A., Yamamoto, K., Kim, S., and Takai, Y. *J. Biol. Chem.*, **264**, 1877 (1989).
10. Kawata, M., Kawahara, Y., Araki, S., Sunako, M., Tsuda, T., Fukuzaki, H., Mizoguchi, A., and Takai, Y. *Biochem. Biophys. Res. Commun.*, **163**, 1418 (1989).
11. Pizon, V., Chardin, P., Lerosey, I., Olofsson, B., and Tavitian, A. *Oncogene*, **3**, 201 (1988).
12. Pizon, V., Lerosey, I., Chardin, P., and Tavitian, A. *Nucleic Acids Res.*, **16**, 7719 (1988).
13. Kitayama, H., Sugimoto, Y., Matsuzaki, T., Ikeda, Y., and Noda, M. *Cell*, **56**, 77 (1989).
14. Kim, S., Mizoguchi, A., Kikuchi, A., and Takai, Y. *Mol. Cell. Biol.*, **10**, 2645 (1990).
15. Mizoguchi, A., Ueda, T., Ikeda, K., Shiku, H., Mizoguchi, H., and Takai, Y. *Mol. Brain Res.*, **5**, 31 (1989).
16. Yamamoto, T., Kaibuchi, K., Mizuno, T., Hiroyoshi, M., Shirataki, H., and Takai, Y. *J. Biol. Chem.*, **265**, 16626 (1990).
17. Yamamoto, J., Kaibuchi, K., and Takai, Y. Manuscript in preparation.
18. Sasaki, T., Kikuchi, A., Araki, S., Hata, Y., Isomura, M., Kuroda, S., and Takai, Y. *J. Biol. Chem.*, **265**, 2333 (1990).
19. Ohga, N., Kikuchi, A., Ueda, T., Yamamoto, J., and Takai, Y. *Biochem. Biophys. Res. Commun.*, **163**, 1523 (1989).
20. Ueda, T., Kikuchi, A., Ohga, N., Yamamoto, J., and Takai, Y. *J. Biol. Chem.*, **265**, 9373 (1990).
21. Fujioka, H., Kikuchi, A., Yoshida, Y., Kuroda, S., and Takai, Y. *Biochem. Biophys. Res. Commun.*, **168**, 1244 (1990).
22. Ohmori, T., Takeyama, Y., Ueda, T., Hiroyoshi, M., Nakanishi, N., Ohyanagi, H., Saitoh, Y., and Takai, Y. *Biochem. Biophys. Res. Commun.*, **169**, 816 (1990).
23. Ueda, T., Kikuchi, A., Ohga, N., Yamamoto, J., and Takai, Y. *Biochem. Biophys. Res. Commun.*, **159**, 1411 (1989).
24. Kikuchi, A., Sasaki, T., Araki, S., Hata, Y., and Takai, Y. *J. Biol. Chem.*, **264**, 9133 (1989).
25. Hata, Y., Kikuchi, A., Sasaki, T., Schaber, M.D., Gibbs, J.B., and Takai, Y. *J. Biol. Chem.*, **265**, 7104 (1990).
26. McCormick, F. *Cell*, **56**, 5 (1990).
27. Hancock, J.F., Magee, A.I., Childs, J.E., and Marshall, C.J. *Cell*, **57**, 1167 (1989).
28. Kawata, M., Farnsworth, C.C., Yoshida, Y., Gelb, M.H., Glomset, J.A., and Takai, Y. *Proc. Natl. Acad. Sci. U.S.A.*, **86**, 8960 (1990).

29. Hiroyoshi, M., Kaibuchi, K., Hata, Y., Kawamura, S., and Takai, Y. *J. Biol. Chem.*, **266**, 2962 (1991).
30. Ballester, R., Furth, M.E., and Rosen, O.M. *J. Biol. Chem.*, **262**, 2688 (1987).
31. Saikumar, P., Ulsh, L.S., Clanton, D.J., Huang, K.-P., and Shih, T.Y. *Oncogene Res.*, **3**, 213 (1988).
32. Kawata, M., Kikuchi, A., Hoshijima, M., Yamamoto, K., Hashimoto, E., Yamamura, H., and Takai, Y. *J. Biol. Chem.*, **264**, 15688 (1989).
33. Hoshijima, M., Kikuchi, A., Kawata, M., Ohmori, T., Hashimoto, E., Yamamura, H., and Takai, Y. *Biochem. Biophys. Res. Commun.*, **157**, 851 (1988).
34. Kawata, M., Kawahara, Y., Sunako, M., Araki, S., Koide, M., Tsuda, T., Fukuzaki, H., and Takai, Y. *Oncogene*, **6**, 841 (1991).
35. Hata, Y., Kaibuchi, K., Kawamura, S., Hiroyoshi, M., Shirataki, H., and Takai, Y. *J. Biol. Chem.*, **266**, 6571 (1991).
36. Frech, M., John, J., Pizon, V., Chardin, P., Tavitian, A., Clark, R., McCormick, F., and Wittinghofer, A. *Science*, **249**, 169 (1990).
37. Molloy, C.J., Bottaro, D.P., Fleming, T.P., Marshall, M.S., Gibbs, J.B., and Aaronson, S.A. *Nature*, **342**, 711 (1990).
38. Ellis, C., Moran, M., McCormick, F., and Pawson, T. *Nature*, **343**, 377 (1990).
39. Kaplan, D.R., Morrison, D.K., Wong, G., McCormick, F., and Williams, L.T. *Cell*, **61**, 125 (1990).
40. Takai, Y., Kikkawa, U., Kaibuchi, K., and Nishizuka, Y. *Adv. Cyclic Nucleotide Protein Phosphorylation Res.*, **18**, 119 (1984).

Protein Phosphorylation and Regulation of Calcium Pump from Cardiac Sarcoplasmic Reticulum

MICHIHIKO TADA, YOSHIHIRO KIMURA, AND MAKOTO INUI

Departments of Medicine and Pathophysiology, Osaka University School of Medicine, Osaka 553, Japan

Sarcoplasmic reticulum (SR) of the myocardium plays a pivotal role in intracellular Ca signaling. Ca release through its Ca release channel induces myofibrillar contraction, while Ca uptake by Ca pump ATPase of this membrane leads to myofibrillar relaxation. A series of studies has indicated that Ca pump of cardiac SR is controlled by another protein, termed phospholamban. This protein was first found by us to serve as a substrate for cAMP-dependent protein kinase. We thereafter showed that phosphorylation of phospholamban by cAMP-dependent protein kinase stimulates the Ca-dependent ATPase activity of Ca pump. Thus, phospholamban is now recognized to be operational in maintaining the functional link between the two important intracellular messengers, Ca and cAMP.

This article reviews several features of the phospholamban-Ca ATPase system, which lead to the important notion that a protein-protein interaction between these SR proteins can modulate Ca signaling in cardiac muscle, resulting in control of myocardial contractility.

I. CARDIAC, SLOW-TWITCH, AND SMOOTH MUSCLES EXPRESS THE SAME PHOSPHOLAMBAN GENE

Shortly after our first series of work (*1–3*) defining cAMP-dependent phosphorylation of phospholamban and its potential role in the regulation of Ca pump in cardiac SR, we reported that this protein may also exist in SR of slow-twitch skeletal muscle, though not in fast-twitch

27

TABLE I

Expression and Chromosomal Localization in Human Genes Encoding Sarco(endo)plasmic
Reticulum Proteins (11–17)

Gene		Tissue					Chromo-some
		Fast	Slow	Cardiac	Smooth	Non-muscle	
Ca-ATPase*	S(E)RCA1	+[a,b]	−	−	−	−	16
	S(E)RCA2	−	+[a]	+[a]	+[b]	+[b]	12
	S(E)RCA3	−	+	+	+	+	
Phospholamban		−	+	+	+	−	6

*Nomenclature based upon Burk et al. (17). [a] and [b] indicates expression of alternatively
spliced forms.

muscle (4). Later, histochemical studies using anti-phospholamban antibodies showed that smooth muscle SR also contains phospholamban (5).
Canine cardiac phospholamban was purified (6, 7) and its primary
structure was determined by both amino acid and cDNA sequencing (8–
10). Using a cDNA fragment as a probe, Northern blot analysis demonstrated that cardiac and slow-twitch muscles in rabbit express the same
phospholamban gene (11). The same gene is expressed in smooth muscle,
but not in non-muscle tissues examined including brain, liver, and kidney
(Table I). As of today, only one phospholamban gene has been identified,
which is located on human chromosome 6 (12). Table I also shows that
there are at least three kinds of Ca ATPase genes expressed in sarco-
(endo)plasmic reticulum. This gene nomenclature is based on Burk et al.
(17). Type 1 gene, SERCA1, is expressed in fast-twitch skeletal muscle
(13). Type 2 gene, SERCA2, encodes two alternatively spliced products,
SERCA2a and SERCA2b. SERCA2a is expressed in slow-twitch skeletal
muscle and cardiac muscle, while SERCA2b is expressed in smooth
muscle and non-muscle tissues (14–16). Type 3 gene, SERCA3, was
found to be expressed in a broad variety of both muscle and non-muscle
tissues. As shown below, the ATPase enzymes expressed by type 2 gene
exhibit a unique molecular and functional interaction with phospholam-

Human	MEKVQYLTRSAIRRASTIEMPQQARQKLQNLFINFCLILICLLLICIIVMLL
Rabbit	*E****L****I*******MPQ****N**N**I******************
Dog	*D****L****I*******MPQ****N**N**I******************
Pig	*D****L****I*******MPQ****N**N**I******************
Chicken	*E****I****L*******VNP****R**E**V******************

Fig. 1. Comparison of amino acid sequences of phospholamban monomer
among human, dog, rabbit, pig, and chicken (10–12, 18, 19). Residues are
represented by the one-letter code. Identical residues among these five species are
shown as asterisks except human sequence.

ban. In Fig. 1, the amino acid sequences of phospholamban deduced from cDNA nucleotide sequence are compared among several species including human (10-12, 18, 19). It is striking that the 19 amino acid residues in the C-terminal portion are completely conserved. It is interesting to note that human phospholamban sequence differs from other species at position 27 (Lys for Asn) and from dog and pig at position 2 (Glu for Asp). Human phospholamban is thus more basic than those from other species.

II. MOLECULAR STRUCTURE OF PHOSPHOLAMBAN

Early observation indicated that phospholamban, either in membrane-bound state or in purified form, exhibits an oligomer structure (20, 21). Even in the presence of detergent such as sodium dodecyl sulfate (SDS), the phospholamban protein still remains an oligomer at temperatures below 50°C. Temperature-dependent conversion between oligomer and monomer was actually demonstrated (20). The molecular weight of phospholamban oligomer was assumed to be 27 kDa, and that of monomer to be 6 kDa based on the mobility on SDS-polyacrylamide gel of Laemmli's system (21). Phosphorylation decreased the mobility of phospholamban on SDS polyacrylamide gels. The patterns of temperature-dependent and phosphorylation-dependent mobility shift suggest that phospholamban is a pentamer comprised of five identical monomers (21). Application of the low-angle laser light scattering technique led us to the conclusion that the molecular weight of phospholamban oligomer is 30,400 Da and that phospholamban is a pentamer (22).

The amino acid sequence of phospholamban deduced from the nucleotide sequence is 52 amino acids long. The calculated molecular weight for the phospholamban monomer is 6,080 (8, 10). Hydropathy plots suggested that phospholamban is an amphipathic peptide; the N-terminal half (Met 1 to Asn 30, called domain I) is hydrophilic, whereas the C-terminal half (Leu 31 to Leu 52, domain II) is extremely hydrophobic. As shown in Fig. 2, domain I is exposed at the cytoplasmic surface, while domain II is embedded within the SR membrane. Analysis of predicted secondary structure indicated that this molecule is rich in alpha-helix; two helices (domains IA and II) are connected by less structured domain IB. Secondary structure analysis using circular dichroism supports this prediction (25).

Domain I was demonstrated to contain the specific phosphorylation

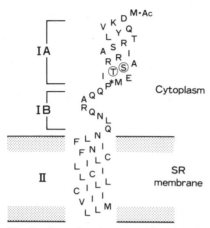

Fig. 2. Secondary structure of canine phospholamban monomer. The two alpha-helices, domain IA and domain II are connected by domain IB, which forms random structure. Domain I is exposed at the cytoplasmic surface, while domain II is anchored in the SR membrane. The circled residues, S and T, represent Ser 16 and Thr 17 which are phosphorylated by cAMP-dependent and Ca/calmodulin-dependent protein kinases, respectively. Asterisk indicates the helix-braking Pro 21.

sites; cAMP-dependent protein kinase phosphorylates Ser 16 and Ca/calmodulin-dependent protein kinase phosphorylates Thr 17 (9, 24). Two Arg residues (Arg 13 and 14) adjacent to the two phosphorylatable residues were proven to be essential for phosphorylation (24), in accord with the consensus sequence for protein kinase substrate.

The molecular property of phospholamban which prohibits detergents such as SDS from penetrating into the oligomer structure has been extensively investigated. The ability to maintain an oligomeric organization would reside in the intramembranous portion since the tryptic fragment devoid of phosphorylation sites remained pentameric (23). The amino acid residues which are essential for oligomeric assembly were identified by site-directed mutagenesis techniques (24). Replacement of one or more of three Cys residues which lie at every five residues in domain II (Cys 36, 41, 46), reduced the stability of phospholamban oligomer; the mutants dissociated to protomers at lower temperature than the wild type oligomer (Fig. 3), indicating that Cys residues, particularly Cys 36, are important for stabilizing the oligomeric structure. It is unlikely, however, that these Cys residues in the neighboring monomers could form disulfide bonds for oligomer formation, because these Cys residues exist as free SH groups (9). Instead, formation of hydrogen

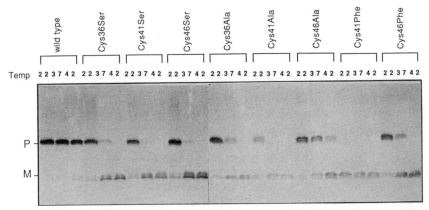

Fig. 3.   Thermal stability of phospholamban mutated in transmembrane cysteine residues (24). The microsomal fractions from COS-1 cells transiently expressing phospholamban proteins mutant in cysteine residues were preincubated with SDS-PAGE loading buffer at 22, 37, and 4°C for 2 min, and then separated in 13.5% SDS-PAGE. They were immunoblotted with monoclonal antibody A1. Temp, preincubation temperature; M, monomeric; and P, pentameric forms of phospholamban.

bonds between neighboring Cys residues is likely responsible for pentamer structure. The fact that the mutant in which Cys was replaced by Ala does exist as oligomer at ambient temperature supports this idea. Probably the size, hydrophobicity, and polarity of the side chain of Cys residues exactly matches the microenvironment, thus allowing the hydrophobic residues in neighboring helices to create optimal stabilizing forces.

III.   EFFECTS OF PHOSPHOLAMBAN PHOSPHORYLATION ON THE REACTION SEQUENCE OF SR Ca PUMP ATPase

Ca pump ATPase is a P-type ATPase which undergoes a complex series of intermediate reaction steps, involving the sequential formation and degradation of phosphorylated intermediates (EP) (27). The ATPase enzyme was found to undergo distinct conformational changes during the transport cycle. The enzyme could be in two different conformational states, E_1 and E_2, which exhibit different affinities for Ca (28). The former has high affinity for Ca, and the latter has low affinity for Ca. Analogous with the substrate-free enzyme, the phosphorylated intermediate EP was also shown to exhibit two comparable forms, E_1P and E_2P. In the physiological state, ATPase translocates Ca according to the reaction sequence shown in Eq. (1):

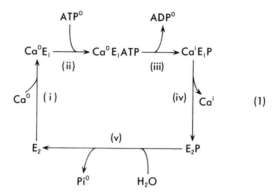

$$\text{(1)}$$

where i and o refer to the inside- and outside-oriented configuration of SR membrane, respectively. In this reaction, conversions of E_2 to E_1 and E_1P to E_2P (steps i and iv) are rate limiting steps.

Phosphorylation of phospholamban enhances the ATPase activity and Ca uptake activity by increasing the Ca affinity of Ca pump ATPase (3). This activation of ATPase reaction results from the acceleration of the turnover of the enzyme reaction (29). Presteady-state kinetic studies revealed that the two rate-limiting steps, *i.e.* steps i and iv, are accelerated by phospholamban phosphorylation (30, 31). These correspond to the conformational transition steps of the ATPase enzyme, suggesting that phospholamban exerts its action by regulating the cation-induced conformational change of the ATPase molecule. A direct protein-protein interaction has been proposed, with the assumption that the conformational state of a region of the ATPase molecule appears to be under the direct control of phospholamban (28, 32).

IV.   INTERMOLECULAR INTERACTION BETWEEN SR Ca PUMP ATPase AND PHOSPHOLAMBAN

An experiment using a crosslinking agent demonstrated the direct interaction between the two SR proteins, Ca pump ATPase and phospholamban (33). The lysine residue of purified phospholamban (Lys 3) was conjugated with Denny-Jaffe crosslinking reagent. Light activation of conjugated phospholamban incubated with purified Ca pump ATPase from cardiac muscle resulted in the formation of a complex only when phospholamban was in the unphosphorylated state and the Ca ATPase was in the Ca-free state (E_2 conformation). The domain of the ATPase which interacts with phospholamban was identified by sequencing the photoaffinity labeled peptide. This peptide, whose two lysine residues (Lys 397 and Lys 400) were labeled, was found to originate from a region which is just six amino acids from the C-terminal side of the phosphorylation domain (Fig. 4). The aspartyl residue (Asp 351) in this phosphorylation domain is phosphorylated to form EP during the transport cycle of ATPase. It remains to be examined which particular residue(s) is essential for the binding. Interestingly, ATPase from fast-twitch skeletal muscle was also shown to contain putative phospholamban binding domain, although phospholamban was not found to exist in SR of this type of muscle (4, 11). Our preliminary data showed that phospholamban can interact with coexpressed SERCA1 Ca ATPase in COS-1 cells. The amino acid sequence in the phosphorylation site among P-type ATPases shares a high degree of homology, whereas the putative phospholamban-binding domain is present only in SR type ATPase, and not in other plasma membrane ATPases (Fig. 4). The finding that phospholamban-binding domain exists in close proximity to the active phosphorylation site of Ca ATPase (Asp 351) indicates that binding of phospholamban to this site could have a significant effect on steps involving phosphorylation of Ca ATPase.

Assuming that phospholamban exhibits its function by binding to a specific region in the ATPase enzyme, it is not yet clear how such a protein-protein interaction could result in significant alterations in the enzymatic activity of Ca pump ATPase, depending upon phosphorylated and unphosphorylated states of phospholamban. It is feasible to propose the following mechanism, based upon our data on enzyme kinetics and chemical observations. The unphosphorylated phospholamban serves to suppress the ATPase activity by binding to the region very close to the phosphorylation site in Ca ATPase (28, 32). Such suppression is relieved when the phosphate incorporation into phospholamban dissociates this protein from ATPase, resulting in the augmentation of the ATPase activity. Experiments using reconstitution techniques (34, 35) or controlled tryptic digestion (36) of phospholamban support this hypothesis.

There are several new observations concerning the mode by which phospholamban interacts with ATPase. These data also lead to the notion that phospholamban would operate to suppress the ATPase. Anti-phospholamban monoclonal antibody A1, the epitope for which was identified to be a region involving Arg 9 (24), was reported to enhance

```
PLB-binding    slow twitch (rabbit)    FILDKVDGETCSLNEFTIIGSTYAP·IGEVHKDDKPVKCHO
peptides:      fast twitch (dog)       FIIDKVDGNLCVLNEFAITGST*APEGEVLKNDKPIRS........

Ca (SR)  1 LGCTSVICSDKTGTLTTNQ MSVCRMFILDKVDGETCSLNEFTIIGSTYAP·IGEVHKDDKPVKCHQTDGLVELATICALCNDSALDYNE  AKGVYEKVGEATE
         2 LGCTSVICSDKTGTLTTNQ MSVCKMFIIDKVDGDFCSLNEFSITGSTYAPEGEVLKNDKPIRSGQFDGLVELATICALCNDSSLDFNE   TKGVYEKVGEATE

Ca (PM)  3 MGNATAICSDKTGTLTMNR MTVVQAYINEK     HY   KKVPEPEAI  PPNILSYL VTGISVNCA YTSKILPEK   EGGLPRHVGNKTE

Na/K     4 LGSTSTICSDKTGTLTQNR MTVAHMWFDNQIHEADTTENQ   SGVSFDKTSATWLALSRIAGLCNRAVFOANQONLPILKRAVAGDASE
         5 LGSTSTICSDKTGTLTQNR MTVAHMWFDNQIHEADTTENQ   SGISFDKTSLSWNALSRIAALCNRAVFQAGQDSVPILKISVAGDASE

H        6 MSGVNMLCSDKTGTLTLNK MEIQEOCFTEEG     NDLKSTLVLAALAAKWREPPRDALDTMVLGAADLDE
         7 LAGVEILCSDKTGTLTKNK LSLHEIYTVEGVS   PDDLMLTAPLAASRKKGLDAIDKAFLKSLKQYPKAK

K        8 AGDVDVLLLDKTGTITLGN RQASEFIPAQGVD   EKTLADAAQLASLADETPEGRSIVILAKQRFNLRER
         9 ANDLDVIMLDKTGTLTQGK FTVTGIEILD      EAYQEEIILKYIGALEAHANHPLAIGIM NYLKEK

           PHOSPHORYLATION DOMAIN
```

Fig. 4. The amino acid sequence of the crosslinked peptides obtained from rabbit cardiac SR ATPase and from canine fast skeletal muscle SR ATPase is shown aligned with homologous regions in other ATPases (33). Phospholamban was conjugated with 125I labeled Denny-Jaffe reagent. Light activation of conjugated phospholamban incubated with purified ATPase resulted in the formation of a complex. The phospholamban-ATPase complex was cleaved at azo linkage using sodium dithionite. This cleavage leaves the 125I label attached to the domain that interacts with phospholamban. The ATPase was then digested with CNBr and fractionated using reverse-phase HPLC followed by ion exchange column chromatography. The 125I-labeled peptides were sequenced. Two dotted lysine residues (K) indicate the sites where the crosslinking reagent bound. 1, rabbit slow-twitch/cardiac muscle SR Ca-ATPase; 2, rabbit fast-twitch muscle SR Ca-ATPase; 3, human plasma membrane Ca-ATPase; 4, sheep plasma membrane Na/K ATPase; 5, torpedo plasma membrane Na/K ATPase; 6, *Saccharomyces cerevisiae* H-ATPase; 7, *Neurospora crassa* H-ATPase; 8, *Escherichia coli* K-ATPase; 9, *Streptococcus faecalis* K-ATPase.

the Ca uptake of cardiac SR vesicles (*37*). The antibody activated the Ca uptake and ATPase activity by enhancing the affinity for Ca without altering the stoichiometry between Ca and ATP (*38*). These effects were comparable to those by the dual phosphorylation of phospholamban by cAMP-dependent and Ca/calmodulin dependent protein kinases. Phospholamban coexpressed with slow/cardiac Ca ATPase in COS cells decreased the affinity of the ATPase for Ca (*39*). Recently the synthesized phospholamban peptides were reconstituted into liposomes with the purified Ca pump ATPase (*35*). The peptide corresponding to 25 amino acid residues from the N-terminus (Met 1 to Arg 25) inhibited Ca uptake and its inhibition was diminished by phosphorylation of the peptide. Our recent experiments have shown that the synthetic peptide corresponding to domain I of phospholamban (Met 1 to Asn 30) inhibited purified cardiac SR Ca pump ATPase activity in a dose-dependent manner; the inhibition was diminished by phosphorylation of the peptide. However, the affinity of the ATPase activity for Ca was not changed by these maneuvers (Table II) (*40*). In other words, this peptide affects only V_{max} of the Ca ATPase activity. Very interestingly, peptides containing domain II, the intramembranous portion of phospholamban, decreased the Ca affinity of the ATPase with the phosphorylation of the peptide, relieving this inhibitor effect (Table II) (*40*). These results are consistent with our hypothesis that unphosphorylated phospholamban acts as a suppressor of Ca pump ATPase and that phosphorylation of phospholamban desuppresses its inhibitory effect. Our present results also indicate that, in terms of phospholamban-mediated regulation of Ca pump ATPase, not only the cytoplasmic domain but also the intramembranous portion of phospholamban are required. A conformational change of domain I caused by the change in electrostatic forces by incorporation of phosphate into phospholamban or steric hindrance brought by the antibody would be transmitted to the transmembrane domain of phospholamban, leading to the dissociation of phospholamban with Ca pump ATPase.

Taking these kinetical and chemical properties into consideration, we might predict the mechanism of the regulation of SR Ca pump ATPase by phospholamban. Domain I of unphosphorylated phospholamban may interact directly with the E_2 form of the ATPase to inhibit the conversion between E_2 and E_1. The interaction is diminished by phosphorylation of phospholamban. Changes in electrostatic properties of the SR membrane on phospholamban phosphorylation may also contribute

TABLE II

Effects of Phospholamban Peptides on Reconstituted Cardiac SR Ca Pump ATPase Activity (*40*)

	n	V_{max} (nmol/mg·min)	K_{Ca} (μM)
Control	4	653.0± 28.3	0.49±0.02
+PLN 1–31 [330]	4	420.2± 12.4*	0.51±0.05
Control	3	634.5±112.4	0.52±0.02
+PLN 28–47 [100]	3	675.7± 43.5	1.33±0.30*
Control	4	637.8± 63.6	0.51±0.04
+PLB 8–47 [100]	3	625.3± 51.8	1.18±0.20*
+PLN 8–47 −P [100]	3	669.0± 68.4	0.72±0.09

Purified canine cardiac Ca pump ATPase in leaky vesicles of phospholipid was preincubated with synthetic partial phospholamban peptide corresponding to the N-terminal hydrophilic domain (PLN 1–31), or was incorporated into liposomes with synthetic phospholamban peptides containing C-terminal hydrophobic sequence (PLN 28–47) by the freeze-thaw-sonication method before being subjected to ATPase assay. For phosphorylation of PLN 8–47, liposomes containing the peptide were first phosphorylated by cAMP-dependent protein kinase, then the phosphorylated vesicles were collected by ultracentrifugation. The molar ratio between the peptide and ATPase is indicated in brackets. V_{max} and K_{Ca} was determined from Ca-dependent profiles of the ATPase activity using the double reciprocal plot of Lineweaver and Burk. Date are mean±S.D. *$p<0.05$ vs. control by unpaired t-test.

to the increased Ca affinity of the ATPase (*41, 42*). Domain II of phospholamban plays a role in the interaction between the two proteins in addition to domain I for two reasons: 1) the domain is well conserved among mammalian and avian species, and 2) synthetic peptide corresponding to domain I of phospholamban did not mimic phospholamban function completely; domain II was necessary for the shift in Ca affinity as mentioned above. It remains to be determined how transmembrane helices of Ca pump ATPase and domain II of phospholamban interact and how the signal of phosphorylation and dephosphorylation of domain I is transduced into the changes in the interaction between domain II of phospholamban and the intramembranous helices of Ca pump ATPase. It is also intriguing to know how the oligomeric structure of phospholamban contributes to the intermolecular interaction between the two molecules. Studies using synthetic phospholamban variants corresponding to the intramembranous portion or biologically synthesized mutants could give us clues to this. Furthermore, mutations on key domain and other related regions of the two proteins may give us the ultimate answer.

SUMMARY

Ca signaling in myocardial cells is regulated by two membrane systems, sarcolemma and SR. The Ca pump ATPase of SR transports Ca into the lumen of SR, lowering the free Ca concentration of the myoplasm. In cardiac SR, Ca pump ATPase has a unique regulatory system by phosphorylation of another cardiac SR protein, phospholamban. This regulatory system *via* cAMP-dependent phosphorylation of phospholamban is thought to be one of the molecular mechanisms of catecholamine's action on myocardium. Phospholamban is a product of a single gene. It is expressed in SR of cardiac, slow-twitch skeletal, and smooth muscles where SERCA2 type Ca pump ATPase is expressed. Phospholamban consists of five monomers with molecular weight of 6,080 daltons. The monomer is an amphipathic polypeptide with C-terminal transmembrane and N-terminal cytoplasmic domains. The cytoplasmic domain has phosphorylation sites catalyzed by cAMP- and calmodulin-dependent protein kinases. One line of evidence indicates that phospholamban acts as a suppressor of Ca pump ATPase, and that its phosphorylation desuppresses the inhibiting effects. A direct interaction between the two proteins was shown by a chemical crosslinking experiment in which the cytoplasmic domain of phospholamban was crosslinked with the region of Ca pump ATPase just C-terminal to the active site of the ATPase. Our recent studies with synthetic peptides of phospholamban further revealed that phospholamban inhibits the Ca pump ATPase at two sites in different manners, the cytoplasmic domain for V_{max} and the intramembrane domain for K_{Ca}. Phosphorylation of phospholamban desuppresses these two inhibitory effects. Thus, phospholamban acts as a regulator of Ca pump ATPase in cardiac SR by a direct protein-protein interaction between the two proteins.

REFERENCES

1. Kirchberger, M.A., Tada, M., and Katz, A.M. *J. Biol. Chem.*, **249**, 6166 (1974).
2. Tada, M., Kirchberger, M.A., and Katz, A.M. *J. Biol. Chem.*, **249**, 6174 (1974).
3. Tada, M., Kirchberger, M.A., and Katz, A.M. *J. Biol. Chem.*, **250**, 2640 (1975).
4. Kirchberger, M.A. and Tada, M. *J. Biol. Chem.*, **251**, 725 (1976).
5. Jorgensen, A.O. and Jones, L.R. *J. Biol. Chem.*, **261**, 3775 (1986).
6. Inui, M., Kadoma, M., and Tada, M. *J. Biol. Chem.*, **260**, 3708 (1985).
7. Jones, L.R., Simmerman, H.K., Wilson, W.W., Gurd, F.R., and Wegener, A.D. *J. Biol.*

   Chem., **260**, 7721 (1985).
 8. Fujii, J., Kadoma, M., Tada, M., Toda, H., and Sakiyama, F. *Biochem. Biophys. Res. Commun.*, **138**, 1044 (1986).
 9. Simmerman, H.K.B., Collins, J.H., Theibert, J.L., Wegener, A.D., and Jones, L.R. *J. Biol. Chem.*, **261**, 13333 (1986).
10. Fujii, J., Ueno, A., Kitano, K., Tanaka, S., Kadoma, M., and Tada, M. *J. Clin. Invest.*, **79**, 301 (1987).
11. Fujii, J., Lytton, J., Tada, M., and MacLennan, D.H. *FEBS Lett.*, **227**, 51 (1988).
12. Fujii, J., Zarin-Herzberg, A., Willard, H.F., Tada, M., and MacLennan, D.H. *J. Biol. Chem.*, **266**, 11669 (1991).
13. Brandl, C.J., deLeon, S., Martin, D.R., and MacLennan, D.H. *J. Biol. Chem.*, **262**, 3768 (1987).
14. Lytton, J. and MacLennan, D.H. *J. Biol. Chem.*, **263**, 15024 (1988).
15. Gunteski-Hamblim, A.-M., Greeb, J., and Shull, G.E. *J. Biol. Chem.*, **263**, 15032 (1988).
16. Lytton, J., Zarin-Herzberg, A., Periasamy, M., and MacLennan, D.H. *J. Biol. Chem.*, **264**, 7059 (1989).
17. Burk, S.E., Lytton, J., MacLennan, D.H., and Shull, G.E. *J. Biol. Chem.*, **264**, 18561 (1989).
18. Verboomen, H., Wuytack, F., Eggermont, J.A., De Jaegere, S., Missiaen, L., Raeymaekers, L., and Casteels, R. *Biochem. J.*, **262**, 353 (1989).
19. Toyofuku, T. and Zak, R. *J. Biol. Chem.*, **266**, 5375 (1991).
20. Le Peuch, C.J., Haiech, J., and Demaille, J.G. *Biochemistry*, **18**, 5150 (1979).
21. Wegener, A.D. and Jones, L.R. *J. Biol. Chem.*, **259**, 1834 (1984).
22. Watanabe, Y., Kijima, Y., Kadoma, M., Tada, M., and Takagi, T. *J. Biochem.*, **110**, 40 (1991).
23. Wegener, A.D., Simmerman, H.K.B., Liepnieks, J., and Jones, L.R. *J. Biol. Chem.*, **261**, 5154 (1986).
24. Fujii, J., Maruyama, K., Tada, M., and MacLennan, D.H. *J. Biol. Chem.*, **264**, 12950 (1989).
25. Simmerman, H.K.B., Lovelace, D.E., and Jones, L.R. *Biochim. Biophys. Acta*, **997**, 322 (1989).
26. Tada, M., Kadoma, M., Inui, M., and Fujii, J. *Methods Enzymol.*, **157**, 107 (1988).
27. Tada, M., Yamamoto, T., and Tonomura, Y. *Physiol. Rev.*, **58**, 1 (1978).
28. Tada, M. and Katz, A.M. *Annu. Rev. Physiol.*, **44**, 401 (1982).
29. Tada, M., Ohmori, F., Yamada, M., and Abe, H. *J. Biol. Chem.*, **254**, 319 (1979).
30. Tada, M., Yamada, M., Ohmori, F., Kuzuya, T., Inui, M., and Abe, H. *J. Biol. Chem.*, **255**, 1985 (1980).
31. Tada, M., Yamada, M., Kadoma, M., Inui, M., and Ohmori, F. *Mol. Cell. Biochem.*, **46**, 73 (1982).
32. Hicks, M.J., Shigekawa, M., and Katz, A.M. *Circ. Res.*, **44**, 384 (1979).
33. James, P., Inui, M., Tada, M., and Carafoli, E. *Nature*, **342**, 90 (1989).
34. Inui, M., Chamberlain, B.K., Saito, A., and Fleischer, S. *J. Biol. Chem.*, **261**, 1794 (1986).
35. Kim, H.W., Steenaart, N.A.E., Ferguson, D.G., and Kranias, E.G. *J. Biol. Chem.*, **265**, 1702 (1990).
36. Kirchberger, M.A., Borchman, D., and Kasinathan, C. *Biochemistry*, **25**, 5484 (1986).
37. Suzuki, T. and Wang, J.H. *J. Biol. Chem.*, **261**, 7018 (1986).

38. Kimura, Y., Inui, M., Kadoma, M., Kijima, Y., Sasaki, T., and Tada, M. *J. Mol. Cell. Cardiol.*, **23**, 1223 (1991).
39. Fujii, J., Maruyama, K., Tada, M., and MacLennan, D.H. *FEBS Lett.*, **273**, 232 (1990).
40. Sasaki, T., Inui, M., Kimura, Y., and Tada, M. *J. Biol. Chem.*, **267**, 1674 (1992).
41. Xu, Z.-C. and Kirchberger, M.A. *J. Biol. Chem.*, **164**, 16644 (1989).
42. Chiesi, M. and Schwaller, R. *FEBS Lett.*, **244**, 241 (1989).

Phosphorylation of Smooth Muscle Myosin: Properties of Myosin Light Chain Kinase

MASAAKI ITO,[*1] VINCE GUERRIERO, JR.,[*2] AND
DAVID J. HARTSHORNE[*2]

*The First Medical Clinic, Mie University Hospital, Tsu, Mie 514, Japan[*1] and Muscle Biology Group, Animal Sciences Department, University of Arizona, Tucson, AZ 85721, U.S.A.[*2]*

Contraction in smooth muscle, as in skeletal and cardiac muscle is initiated by an increase in intracellular Ca^{2+}. The Ca^{2+} transients involved in all muscle types are similar, and the concentration of free Ca^{2+} required to cause contraction is about 0.5 μM. The Ca^{2+} targets in these muscle types are, however, distinct. In striated muscle troponin C serves as the Ca^{2+} receptor and its interaction with Ca^{2+} induces a sequence of conformational changes in the thin filament proteins that allows interaction of the myosin cross-bridge with actin. In smooth muscle troponin is absent and a different regulatory scheme is dominant. This involves the phosphorylation of myosin.

Myosin in relaxed muscle is in the dephosphorylated state and in an "inactive" conformation. Phosphorylation of each of the two 20,000-dalton light chains causes an increase in the actin-activated ATPase activity of myosin and this increased enzymatic activity is thought to equate with the development of tension in intact smooth muscle cells. Myosin is phosphorylated by myosin light chain kinase (MLCK) and a critical feature of the regulatory mechanism is the coupling of kinase activation to the formation of the Ca^{2+}-calmodulin (CaM) complex. Thus, in smooth muscle the Ca^{2+} target is CaM and since its affinity for Ca^{2+} is similar to troponin C both striated and smooth muscles operate over similar ranges of Ca^{2+} concentrations.

According to the simplest interpretation of the phosphorylation theory, phosphorylation of myosin leads to an "active state" and the

41

development of tension. From mechanics of skeletal muscle contraction it is assumed that tension is a reflection of the number of cross-bridges acting in parallel. Thus, theoretically an increase in the level of phosphorylation of myosin should lead to an increase in tension (more cross-bridges being recruited) but at constant velocity. In practice this relationship is not uniformly realized. There are many reports in which tension can be maintained at low phosphorylation levels (the latch state) and also reports in which low tension is accompanied by relatively high phosphorylation levels. These deviations from the simplest interpretation of the phosphorylation theory have led to the suggestion that alternative, or complementary, regulatory mechanisms may exist. Candidates here include the thin filament proteins, caldesmon, and/or calponin. Another possibility is that the phosphorylation mechanism is not as simple as depicted. Different states or conformations of myosin may occur during the cross-bridge cycle, or as a result of dephosphorylation, and these may provide the necessary flexibility to the phosphorylation scheme.

It is generally agreed, however, that phosphorylation of myosin is required to initiate contraction in smooth muscle (1). During the onset of contraction there is usually a reasonable correlation between tension development and myosin phosphorylation and deviations from this behavior occur during subsequent phases of contraction. In relaxed muscle the level of myosin phosphorylation is low and thus dephosphorylation is required for relaxation. Again, the details are not established and the identity of the phosphatase involved is not universally accepted. From studies done with phosphatase inhibitors, calyculin-A and okadaic acid, it was suggested (2) that the phosphatase acting on myosin was a type-1 enzyme, although this needs to be confirmed by additional studies. One of the most important points to clarify is whether phosphatase activity is regulated. It is now assumed, in the absence of evidence to the contrary, that phosphatase activity is constant in the contracted and relaxed states. From the combined action of MLCK plus the phosphatase this would lead to a "pseudo-ATPase" and thus constitute a futile energy drain. Intuitively, this seems unsatisfactory and regulated phosphatase activity is more reasonable.

Smooth muscle myosin in the monomeric state can exist in either of two conformations, a folded (10S) or an extended (6S) form. Increasing ionic strength and phosphorylation favor the transition from 10S to 6S. Accompanying the conformational transition is an alteration of myosin ATPase activity and the idea evolved that the conformation of myosin (*i.e.*, either 6S or 10S) determined enzymatic activity (*3*). Thus the increase of actin-activated ATPase activity may result from a conformational change of myosin. Since these conformational transitions are poised delicately at physiological ionic strength it was suggested (*3*) that similar changes might occur *in vivo*. However, in relaxed and contracting smooth muscle, myosin exists in thick filaments (*4*) and it is therefore difficult to imagine that the entire 10S-6S transition could occur. Possibly only certain parts of the transition are allowed in filamentous myosin and several studies were directed towards identifying these critical changes. One approach was to use limited proteolysis as a probe of myosin conformation. It was found (*5*) that the subfragment 1 (S1)-subfragment 2 (S2) junction is altered during the 10S-6S transition. This led to the suggestion (*5, 6*) that this region of the molecule, site B, is involved in expression of enzymatic activity. It was proposed that in the 10S state the myosin heads are constrained and held in an inactive configuration. Phosphorylation of the 20,000-dalton light chain, thought to be in proximity to S2 and the head-neck junction, alters flexibility of S2 and allows interaction of the myosin head(s) with actin. Based on electron microscopy and hydrodynamic measurements of phosphorylated and dephosphorylated heavy meromyosin (HMM), Suzuki *et al.* (*7*) also suggested that the orientation of the myosin heads could play an important role in determining enzymatic activity.

In addition to the S1-S2 site a second site, site A, sensitive to conformation was detected (*6*) approximately 68-kD from the N-terminus of the heavy chain. This site was masked in the 10S conformation and was also blocked by binding of actin to 6S myosin (*8*). It was assumed, therefore, that site A is at or close to the actin-binding site and, if so, then binding of actin to 10S myosin should be reduced compared to 6S myosin. This was demonstrated (*8*) and in the presence of AMPPNP the apparent dissociation constant for dephosphorylated myosin was about 40 μM compared to 0.3 μM for phosphorylated myosin. Site A is located at the junction of the central and C-terminal domains of S1, and in skeletal muscle myosin this is also the approximate location of the actin-binding site. The role, if any, of site A in the contractile mechanism is not known. It is interesting to speculate, however, that dephosphorylation of the attached cross-bridge (actin-myosin complex) would allow a conformational change at site B, the S1-S2 junction, but would not allow changes at the occupied actin-binding site. This state could be important if its subsequent dissociation rate is slow compared to the phosphorylated cycling cross-bridge, and may account for the maintenance of tension at reduced velocity. An additional and attractive feature of this hypothesis is that since the affinity for actin of the dephosphorylated myosin is reduced it would not reattach and phosphorylation *via* MLCK would be required for the attachment to actin to occur.

I. MYOSIN LIGHT CHAIN KINASE

Although details of the regulatory mechanism in smooth muscle are not available, it is clear that MLCK and the myosin light chain phosphatase play critical roles. Of these key enzymes considerably more is known about MLCK.

1. Distribution

MLCK plays a dominant role in the regulation of vertebrate smooth muscle activity and probably also in several motile processes of non-muscle cells. It has been identified in a variety of eucaryotic cells including fibroblasts, macrophages, lymphocytes, intestinal epithelial brush border cells, teleost retinal cones, and thyroid. This is only a partial listing, but it clearly underlines the importance of myosin phosphorylation in several cell processes. It has been suggested that phosphorylation of myosin alters its state of aggregation (*9, 10*) and this may be particularly important in non-muscle cells. In the "resting" cell it is proposed that myosin exists largely as the soluble 10S monomer; phosphorylation of myosin would then promote the 10S-6S transition and subsequent formation of myosin aggregates. (It is assumed that the latter is the form in which myosin functions in the relevant process). MLCK is also found in striated muscle and, although it does not exert a dominant role, myosin phosphorylation is thought to modify contractile response under conditions of sub-maximal stimulation (*11*).

The subcellular localization of MLCK is not established. It binds to both myosin (*12*) and actin (*13*) but may not be restricted under all conditions to either the thin or thick filaments. An interesting point is that if the concentration of MLCK in smooth muscle is about 4 μM, then there would be approximately 2 molecules of MLCK per thin filament. If binding to F-actin is not random then the only unique binding sites on the thin filament would be at either end. Since one end of the thin filament is attached to the dense body the free end of the actin filament is a potential location for MLCK. In smooth muscle there is about 100

μM 20,000-dalton light chain, this gives a substrate : enzyme ratio of about 25 : 1, or about 1 MLCK molecule per 12 myosins. If MLCK is bound to the thick filament it must have some mechanism to move from one myosin to the next, and also it should be able to recognize only those myosins capable of force production. (Obviously myosin in a non-overlap zone is incapable of tension development). Intuitively it is more reasonable to anchor MLCK to the thin filament and to phosphorylate only those myosins in its vicinity.

2. Domains of MLCK

Several studies have been carried out to define various regions of the MLCK molecule and various models have been proposed. In general, these depict an N-terminal region of variable length, a central active site, and the C-terminal region containing the CaM-binding site (see ref. 1). Identification of the residues involved in each segment was facilitated by sequences derived for gizzard MLCK from a cloned partial cDNA (14) and, recently, from the complete cDNA (15). In addition to the smooth muscle MLCK, complete sequences are also available for the skeletal muscle MLCK from rabbit (16) and rat (17).

Using the residue assignments from the sequence of the complete cDNA (15) the following domains may be defined: the active site spans residues D517 to R762, with the consensus ATP binding sequence (G-X-G-X-X-G-) at its N-terminal end beginning at G526; the CaM-binding site is located between residues A796 to S815. At the end of this sequence is one of the residues, S814, phosphorylated by the cAMP-dependent protein kinase. Other areas of the molecule are less well defined, for example, preliminary results (Kanoh and Hartshorne, unpublished) suggest that the actin-binding domain is at the N-terminal end of the molecule, but its limits are not defined. It is interesting that the N-terminal part of MLCK is rich in proline and this may generate an extended or asymmetric structure facilitating interaction with the F-actin helix.

An unusual feature of the smooth muscle MLCK is the C-terminal part of the molecule that is synthesized as an independent protein (18). The initiation site is not established but is probably M818, based on the content of methionine in the molecule, i.e., 4 mol/mol. The first residue would therefore be I819 (assuming that M818 is not incorporated) and the molecule would be composed of 153 residues with an approximate molecular weight of 17,600. The corresponding sequence is not present in skeletal muscle MLCK where the C-terminus is close to the end of the CaM-binding site. This protein is therefore not present in either cardiac or skeletal muscle. Since the protein contains the tail of the kinase molecule it has been termed "telokin" (*telos*, Gr. tail).

In the earlier studies of Guerriero *et al.* (*14*) a puzzling feature was that the 2.1-kb DNA (to MLCK) hybridized to two sizes of RNA: a 5.5-kb RNA, thought to be the mRNA for the complete molecule, and a smaller 2.7-kb RNA. The latter was determined to be related to the C-terminal part of the molecule, but its translation product was not detected by a polyclonal antibody to MLCK on Western blots. This was particularly surprising since the smaller mRNA was present at higher concentrations than the 5.5-kb RNA. Subsequently it was realized that if different conditions were used during the transfer to nitrocellulose a second antigen could be detected, and this was telokin. This is an abundant protein in gizzard and, in fact, was first isolated in 1977 as a by-product of the CaM purification procedure (*19*) and was shown not to substitute for CaM in the activation of MLCK.

The evidence to indicate that telokin is expressed independently and is not a proteolytic degradation product of MLCK is as follows: 1) its N-terminal amino acid is blocked; 2) the concentration of telokin is at least $15 \mu M$ in gizzard (compared to about $4 \mu M$ MLCK). This is consistent with the relatively-abundant 2.7-kb RNA; 3) no fragments corresponding to the central and N-terminal regions of MLCK are detected. These would be expected if telokin was generated *via* proteolysis. Thus, although it is not conclusive, it appears likely that the MLCK gene is under the control of two promoters. It is possible that regulation of each is independent since the message for telokin, not MLCK, is hormonally regulated in the oviduct (*20*).

To date the function of telokin is not known. Because of its acidic nature (pI ~ 4.5) and metachromatic staining with Stains-all (*18*) it was suspected to bind cations. However, it does not bind Ca^{2+} with high affinity and Ca^{2+}-binding is competitive with Mg^{2+}. It is possible that its interaction with Ca^{2+} may be related to function but this remains to be determined. In the absence of Mg^{2+}, two classes of binding sites were detected: a higher affinity class of $K_D \sim 540 \mu M$ and an n value of 4 and a lower affinity class of K_D approaching 1 mM. Telokin also contains a phosphorylation site for cAMP-dependent protein kinase, corresponding to residue S828 of the native molecule. Phosphorylation at this site does not influence Ca^{2+}- (or Mg^{2+}-) binding.

3. Autoinhibition of MLCK

One of the intriguing questions with CaM-dependent enzymes in general is why is the apoenzyme inactive? For the skeletal muscle MLCK the lack of activity in the apoenzyme was suggested to be due to inhibition *via* an inhibitory domain (*21*). With the gizzard MLCK this situation may also occur, but it has been suggested that the recognition by the active site of the inhibitory domain is due to the latter's similarity to the light chain substrate, leading to the term pseudosubstrate domain (*22*). In the apoenzyme the active site would be occupied by the pseudosubstrate domain and the binding of Ca^{2+}-CaM would release this interaction. This hypothesis is based on similarities in juxtaposition of the basic residues in both the MLCK sequence and the N-terminal sequence of the 20,000-dalton light chain, where H805 of MLCK is aligned with S19 of the light chain. The pseudosubstrate domain would therefore extend over 19 residues from S787 to H805.

The concept of autoinhibition of a kinase *via* a sequence similar to its substrate is not unique to MLCK and, in fact, was proposed initially for cAMP-dependent protein kinase (review ref. *23*) and has subsequently been proposed for other protein kinases, *i.e.*, protein kinase C (*24*) and CaM-dependent protein kinase II (*25, 26*). Attractive though this concept is, it is not universally accepted for MLCK; and Ikebe *et al.* (*27*) have suggested that the inhibitory domain of gizzard MLCK is contained within the sequence D777 to K793. Obviously this sequence is close to the pseudosubstrate sequence, and in fact, overlaps for several residues, but it is not based on the pseudosubstrate concept, and as such poses a challenge to the latter.

Studies with synthetic peptides have been useful in defining the regions of MLCK involved in inhibition. Peptides spanning the regions A783 to G804 and A483 to Q801 were potent inhibitors of kinase activity, both for the native enzyme and for the constitutively active fragment (*28*). This is not surprising since they contain most of the inhibitory sequence D777 to K793 and most of the pseudosubstrate sequence S787 to H805. Peptides A796 to S815 and A796 to V807 also were inhibitory, but were less effective than the 783 to 804, or 483 to 801 (*28*). In a systematic study using many synthetic peptides Foster *et al.* (*29*) showed that within the sequence L786 to V807 there are five peptide sequences of 4 to 6 amino acids that contribute individually to direct inhibition, as well as being involved in CaM binding. The peptides are: L786 to D789,

R790 to M791, K792 to A796, R797 to Q801, and K802 to V807. Each of these subgroups contains basic amino acids. Residues C-terminal to R808 are involved only in CaM binding. From these data it is tentatively concluded that no single short sequence of MLCK is entirely responsible for autoinhibition and several regions may be involved.

Limited proteolysis of MLCK has been useful in defining areas of the molecule involved in regulation. It was shown (30) that on proteolysis with trypsin the CaM-dependent activity of MLCK decreased to yield eventually an inactive kinase fragment. Continued hydrolysis caused an increase in kinase activity but this was CaM-independent, and reflected the formation of the constitutively-active kinase fragment. Neither the inactive- nor the active fragment bound CaM. These results were interpreted as follows: initially the CaM-binding site was removed by proteolysis but the inhibitory sequence was retained. This generates an unregulated kinase fragment frozen in the inhibited state. The activity of the inactive fragment was determined to be about 0.1% of that of the native enzyme (30). Theoretically this value should and does approach that observed for the apoenzyme; on further proteolysis the inhibitory site is removed and this generates an unregulated kinase fragment frozen in the active state. The activity of the active fragment should equal that of the holoenzyme. In practice, this activity was slightly less than that of the native enzyme plus Ca^{2+}-CaM.

In addition, it was found that neither fragment bound to actin and therefore additional cleavages sites would be predicted towards the N-terminal half of the molecule. The molecular weights of the inactive and active fragments were estimated as 64- and 61-kD, respectively, and the inhibitory region therefore must be contained within a 3-kD span of the enzyme. Obviously this segment is large enough to accommodate both the inhibitory sequence and the pseudosubstrate sequence.

Attempts to further define the autoinhibitory region using other proteases were not entirely successful, although these studies did provide some useful information. It was shown (31) that thermolysin generated only the inactive fragment and this proved useful for its preparation. Endoproteinase Lys-C generated both fragments. (Other proteases were tried but did not offer any obvious advantages). The objective of these studies was to determine the N- and C-terminal boundaries of each fragment and in this way to define limits of the autoinhibitory sequence. It was found (15) that R808 is the C-terminal residue of the tryptic inactive fragment and K779 the C-terminal residue of the active fragment

(*31*). For the other inactive fragments the C-terminal residues were A806 and K802 for thermolysin and endoproteinase Lys-C, respectively. From these results it was concluded that the conversion of the inactive- to active fragments is due entirely to proteolysis on the C-terminal end of MLCK and the inhibitory region is contained within the sequence N780 (tryptic active-fragment) and K802 (endoproteinase Lys-C inactive-fragment). This stretch of 23 residues is too long to discriminate between the inhibitory sequence of Ikebe *et al.* (*27*) and the pseudosubstrate hypothesis.

An additional point can be made from these proteolysis studies. In comparing the conserved sequences of several protein kinases Hank *et al.* (*32*) determined that the C-terminal limit of the active site of MLCK was R762 in subdomain XI. The C-terminal amino acid of the active fragment produced by endoproteinase Lys-C was determined to be K760. This active fragment had no obvious kinetic distinctions from the native enzyme (except for CaM-dependence), and therefore it appears that R762 is not essential for kinase activity nor does it seem to play an important role in determining enzymatic parameters of MLCK.

4. Mutagenesis of MLCK

One method to define more precisely those residues involved in autoinhibition is by application of site-directed mutagenesis. In these preliminary studies only truncated mutants were expressed. The area of the molecule that was focused on was that indicated from the proteolysis experiments, namely to the C-terminal side of K779. This region includes the autoinhibitory sequence, the CaM-binding site and telokin.

The expression vectors that were used were mostly pRIT2T (Pharmacia LKB Biotechnology) and, to a lesser extent, pET-3a (*33*). The advantage of the former is that this vector contains a fragment coding for protein A that has IgG binding ability, followed by the pUC9 polylinker region for the insertion of cDNAs. A cDNA inserted in the correct reading frame and orientation will be expressed as a fusion protein containing the segment of protein A. This facilitates purification of the mutants using a monoclonal IgG_1 affinity column. The purpose of the second expression vector was to determine whether the inclusion of a fragment of protein A markedly altered the biological properties of the mutants. Where possible, the expressed proteins were purified by both CaM-affinity and IgG_1-affinity chromatography and, where mutants lacked either the CaM-binding site or the protein A fragment, only one affinity column was used. The majority of the expressed protein was incorporated into inclusion bodies and was not soluble. Attempts to solubilize and renature this protein have failed. Consequently, the amount of available protein was limited so that only enzymatic parameters for the mutants were determined.

Each mutant was initiated at L447 and therefore the mutants were designated according to their C-terminal amino acid. These were: pMK·Glu 972, pMK·Trp 800, pMK·Lys 793, pMK·Thr 778, and pET·Glu 972.

The two "complete" mutants, pMK·Glu 972 and pET·Glu 972 contained autoinhibitory and CaM-binding domains and were predicted to be CaM-dependent. This was observed and there was no detectable difference between the pMK- and pET-systems, indicating that the protein A fragment did not influence kinase activity. The specific activity of the mutants in the presence of Ca^{2+} and CaM was between 2.2 to 2.6 μmol P_i transferred/min·mg^{-1} (based on the percentage of the pertinent band on sodium dodecyl sulfate (SDS)-electrophoresis). The three truncated mutants each possessed lower kinase activity, with pMK·Trp 800 being considerably lower than the others. For each of the five mutants K_m values for ATP and light chain were constant and were similar to values obtained with the native enzyme. The important distinction between these expressed proteins was that the three truncated mutants were constitutively-active and were not dependent on Ca^{2+}-CaM.

Unfortunately, however, the finite level of kinase activity obtained with the pMK·Trp 800 mutant was too high to allow an unequivocal decision concerning the presence or absence of the autoinhibitory sequence. Theoretically, based on either of the two proposals concerning the inhibitory domain, pMK·Trp 800 should have been inactive. The failure to realize this is not understood but may be due to either the presence of an incomplete inhibitory sequence (*i.e.*, residues C-terminal to W 800 may be involved), incorrect folding of the expressed protein, or to a limited extent of proteolysis occurring during isolation. Whatever the reason, a second procedure was required to evaluate the mutants for the presence of the inhibitory domain. To achieve this we resorted to use of limited proteolysis by trypsin, since there was ample precedence to document that the removal of the inhibitory domain by trypsin resulted in activation of kinase activity. When this procedure was applied to the three truncated mutants, activation was observed in only one instance, that of pMK·Trp 800. The two shorter mutants, pMK·Lys 793 and pMK·Thr 778, were not markedly affected by tryptic hydrolysis. Thus, the conclusion from both the specific activities of the expressed proteins and the effects of tryptic hydrolysis is that an important component of the inhibitory sequence is present only in pMK·Trp 800. The sequence, Y794MARRKW800 must therefore be regarded as critical to the mechanism of autoinhibition. The last four residues of this sequence are probably most important to the inhibitory effect, but this must be validated using additional mutants.

Considering that pMK·Trp 800 possessed low, but measurable activity and the results from the use of synthetic peptides (29) it is possible that the sequence 794 to 800 is only part of the autoinhibitory domain. Other regions C-terminal to this sequence may also be involved and necessary for complete inhibition of the apoenzyme. However, if it is accepted that the sequence 794 to 800 is critical to the inhibitory mechanism then this would argue strongly in favor of the pseudosubstrate hypothesis. The CaM-binding site extends from A796 to S815 and would therefore overlap with the inhibitory sequence by five residues. With this arrangement it is likely that the binding of CaM would release the constraints imposed on the apoenzyme by the sequence Y794 to W800.

SUMMARY

A major regulatory mechanism in smooth muscle is the phosphorylation of myosin. The basic tenets of the theory are that phosphorylation of the two 20,000-dalton light chains "activates" the contractile apparatus by increasing the actin-activated ATPase activity of myosin and that dephosphorylation of the light chains reverses this process. Thus the two key regulatory enzymes in this mechanism are the MLCK and the myosin light chain phosphatase. Considerably more is known about MLCK and this is the focus of this review. Recently there has been considerable interest in the domain structure of MLCK and, in particular, that region of the molecule responsible for inhibition of the apoenzyme, the autoinhibitory site. Studies using limited proteolysis restricted the inhibitory region to the C-terminal end of the molecule to the sequence N780 to K802. This stretch of twenty three amino acids, however, is too long to discriminate between the various hypotheses proposed to account for inhibition and a more selective procedure was required. This was site-directed mutagenesis of MLCK in a bacterial expression system. Each mutant was initiated at L447 but was truncated to different extents at the C-terminal end. The mutants containing the sequences L447 to T778 and K793 were both constitutively-active and therefore did not contain the inhibitory domain. In contrast, the mutant ending at W800 was inhibited. It is suggested that a critical region of the inhibitory domain is contained within the sequence Y794 to W800. These results are consistent with the hypothesis that inhibition of the apoenzyme is due to interaction of the active site of MLCK with a pseudosubstrate sequence.

Acknowledgment

This work was supported by grants HL-43651 (to V.G.) and HL-23615 (to D.J.H.) from the National Institutes of Health, U.S.A.

REFERENCES

1. Hartshorne, D.J. *In* "Physiology of the Gastrointestinal Tract," Vol. I, 2nd Ed., ed. L.R. Johnson, p. 423 (1987). Raven Press, New York.
2. Ishihara, H., Martin, B.L., Brautigan, D.L., Karaki, H., Ozaki, H., Kato, Y., Fusetani, N., Watabe, S., Hashimoto, K., Uemura, D., and Hartshorne, D.J. *Biochem. Biophys. Res. Commun.*, **159**, 871 (1989).
3. Ikebe, M., Hinkins, S., and Hartshorne, D.J. *Biochemistry*, **22**, 4580 (1983).
4. Somlyo, A.V., Butler, T.M., Bond, M., and Somlyo, A.P. *Nature*, **294**, 567 (1981).
5. Ikebe, M. and Hartshorne, D.J. *J. Biol. Chem.*, **259**, 11639 (1984).
6. Ikebe, M. and Hartshorne, D.J. *Biochemistry*, **24**, 2380 (1985).
7. Suzuki, H., Stafford, W.F., III, Slayter, H.S., and Seidel, J.C. *J. Biol. Chem.*, **260**, 14810 (1985).
8. Ikebe, M. and Hartshorne, D.J. *Biochemistry*, **25**, 6177 (1986).
9. Suzuki, H., Onishi, H., Takahashi, K., and Watanabe, S. *J. Biochem.*, **84**, 1529 (1978).
10. Scholey, J.M., Taylor, K.A., and Kendrick-Jones, J. *Nature*, **287**, 233 (1980).
11. Sweeney, H.L. and Stull, J.T. *Proc. Natl. Acad. Sci. U.S.A.*, **87**, 414 (1990).
12. Sellers, J.R. and Pato, M.D. *J. Biol. Chem.* **259**, 7740 (1984).
13. Dabrowska, R., Hinkins, S., Walsh, M.P., and Hartshorne, D.J. *Biochem. Biophys. Res. Commun.*, **107**, 1524 (1982).
14. Guerriero, V., Jr., Russo, M.A., Olson, N.J., Putkey, J.A., and Means, A.R. *Biochemistry*, **25**, 8372 (1986).
15. Olson, N.J., Pearson, R.B., Needleman, D.S., Hurwitz, M.Y., Kemp, B.E., and Means, A.R. *Proc. Natl. Acad. Sci. U.S.A.*, **87**, 2284 (1990).
16. Takio, K., Blementhal, D.K., Walsh, K.A., Titani, K., and Krebs, E.G. *Biochemistry*, **25**, 8049 (1986).
17. Roush, C.L., Kennelly, P.J., Glaccum, M.B., Helfman, D.M., Scott, J.D., and Krebs, E.G. *J. Biol. Chem.*, **263**, 10510 (1988).
18. Ito, M., Dabrowska, R., Guerriero, V., Jr., and Hartshorne, D.J. *J. Biol. Chem.*, **264**, 13971 (1989).

19. Dabrowska, R., Aromatorio, D., Sherry, J.M.F., and Hartshorne, D.J. *Biochem. Biophys. Res. Commun.*, **78**, 1263 (1977).
20. Russo, M.A., Guerriero, V., Jr., and Means, A.R. *Mol. Endocrinol.*, **1**, 60 (1987).
21. Edelmann, A.M., Takio, K., Blumenthal, D.K., Hansen, R.S., Walsh, K.A., Titani, K., and Krebs, E.G. *J. Biol. Chem.*, **260**, 11275 (1985).
22. Kemp, B.E., Pearson, R.B., Guerriero, V., Jr., Bagchi, I., and Means, A.R. *J. Biol. Chem.*, **262**, 2542 (1987).
23. Taylor, S.S. *J. Biol. Chem.*, **264**, 8443 (1989).
24. House, C. and Kemp, B.E. *Science*, **238**, 1726 (1987).
25. Hanley, R.M., Means, A.R., Ono, T., Kemp, B.E., Burgin, K.E., Waxham, N., and Kelly, P.T. *Science*, **237**, 293 (1987).
26. Payne, M.E., Fong, Y.-L., Ono, T., Colbran, R.J., Kemp, B.E., Soderling, T.R., and Means, A.R. *J. Biol. Chem.*, **263**, 7190 (1988).
27. Ikebe, M., Maruta, S., and Reardon, S. *J. Biol. Chem.*, **264**, 6967 (1989).
28. Ikebe, M. *Biochem. Biophys. Res. Commun.*, **168**, 714 (1990).
29. Foster, C.J., Johnston, S.A., Sunday, B., and Gaeta, F.C.A. *Arch. Biochem. Biophys.*, **280**, 397 (1990).
30. Ikebe, M., Stepinska, M., Kemp, B.E., Means, A.R., and Hartshorne, D.J. *J. Biol. Chem.*, **260**, 13828 (1987).
31. Ito, M., Hartshorne, D.J., Pearson, R., and Kemp, B.E. *Biophys. J.*, **55**, 474a (1989).
32. Hank, S.K., Quinn, A.M., and Hunter, T. *Science*, **241**, 42 (1988).
33. Studier, W.S., Rosenberg, A.H., and Dunn, J.J. *Methods Enzymol.*, **185**, 60 (1990).

Role of Ca^{2+}-Binding Protein in Signal Transduction

HIROYOSHI HIDAKA, HIROSHI TOKUMITSU, AND
AKIHIRO MIZUTANI

Department of Pharmacology, Nagoya University School of Medicine, Nagoya 466, Japan

Increasing concentration of intracellular Ca^{2+} plays an important role in many cellular responses through its binding proteins. Ca^{2+}-binding proteins can be classified into several groups on the basis of their properties. One is a well-known group including calmodulin, troponin C, parvalbumin, S-100 protein, and several S-100 related proteins which have an EF-structure in each molecule. Another is a group of γ-carboxy-glutamine containing proteins. The annexin family is a different type of Ca^{2+}-binding protein which interacts with phospholipid in a Ca^{2+} dependent manner.

In this paper, we focus on the Ca^{2+}/calmodulin dependent pathway. Calmodulin is thought to regulate the activity of many enzymes which are involved in smooth muscle contraction, including Ca^{2+}-phosphodiesterase (*1*), adenylate cyclase (*2*), and myosin light chain kinase (*3*). Ca^{2+}/ calmodulin kinase II (CaM kinase II) is also activated by calmodulin which exists in abundance in brain (*4*). Another role of calmodulin is regulation of the function of non-enzyme proteins such as caldesmon, MAP$_2$ (*5*), τ, and spectrin which are all cytoskeletal associated protein. Here, we report on two calmodulin associated proteins: one is the novel calmodulin binding protein (ACAMP-81) and the other is CaM kinase II, whose function we investigated using the newly synthesized inhibitor (KN-62).

I. PURIFICATION AND CHARACTERIZATION OF ACAMP-81 (*6, 7*)

A novel calmodulin binding protein was discovered from bovine brain and called *a*cidic *cal*modulin binding *p*rotein, M.W. of *81* kDa

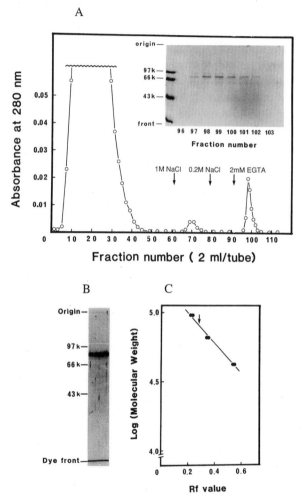

Fig. 1.   Purification of ACAMP-81 by calmodulin-affinity chromatography. A: representative elution profile of calmodulin affinity chromatography of ACAMP-81. ○ absorbance of 280 nm. Inset: 10% SDS-PAGE of eluted fractions. B: Coomassie blue staining of 10% SDS-PAGE of purified ACAMP-81. C: determination of molecular weight of ACAMP-81. Molecular weight marker: upper, phosphorylase b: M_r 97,400; middle, albumin: M_r 66,200; lower, ovalbumin: M_r 42,699. ACAMP-81 is indicated by an arrow.

(ACAMP-81). Purification of ACAMP-81 from Triton X-100 soluble fraction of bovine brain membrane was performed using calmodulin affinity chromatography. Two hundred micrograms of purified ACAMP-81 was obtained from 500 g of bovine brain (Fig. 1). Molecular weight of this protein was estimated to be 81,000. pI value of ACAMP-81 was calculated to be 4.3, showing it to be an acidic protein. Physicochemical properties (Stokes radius = 52 Å, $S_{20w} = 20$) revealed that this protein had a notable elongated shape.

Interaction of calmodulin and other Ca^{2+}-binding proteins with ACAMP-81 was observed by alkaline gel electrophoresis in the presence of 1 M urea. The mobility of ACAMP-81 was not altered in either the presence or absence of Ca^{2+}, however, it was in the presence of Ca^{2+} and calmodulin. The mobility was somewhat reduced in the absence of Ca^{2+}, suggesting that ACAMP-81 formed a complex with calmodulin in a Ca^{2+} dependent manner (Fig. 2). In other Ca^{2+}-binding proteins, S-100 protein

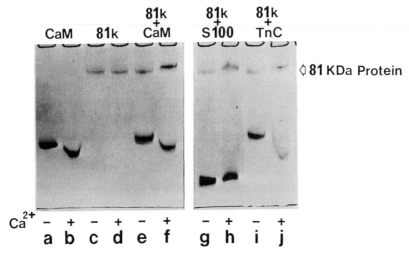

Fig. 2. Interaction of ACAMP-81 with Ca^{2+} binding proteins. Electrophoresis was carried out using 7.5% polyacrylamide gel containing 1 M urea. Samples were incubated at 30°C for 30 min in a solution containing 1 M urea, 20 mM Tris-glycine (pH 8.3) in the presence of 2 mM CaCl$_2$ (lanes b, d, f, h, and j) or in the presence of 2 mM EGTA (lanes a, c, e, g, and i) and subjected to electrophoresis. Gels were stained with Coomassie blue. A: lanes a and b, calmodulin (0.8 μg); c, and d, purified ACAMP-81 (0.5 μg); lanes e and f ACAMP-81 (0.5 μg)+calmodulin (0.8 μg). B: lanes g and h, ACAMP-81 (0.5 μg)+S-100 protein (1.25 μg); lanes i and j, ACAMP-81 (0.5 μg)+troponin C (1 μg).

from brain and troponin C from skeletal muscle, the same results were observed. Stoichiometry of calmodulin binding to ACAMP-81 was measured by the same electrophoretic system and ^{125}I-calmodulin. One mol of ACAMP-81 could bind to an equimolar amount of calmodulin with K_d value of 0.65 μM.

To determine the function of ACAMP-81, we searched for an associated protein using the ^{125}I-ACAMP-81 overlay method, and found the associated protein in high salt extract from bovine brain (Fig. 3); it had a molecular weight of 84 kDa on sodium dodecyl sulfate-polyacrylamide gel electrophoresis (SDS-PAGE). The partially purified ACAMP-81 associated protein detected by ^{125}I-ACAMP binding was a good substrate for cAMP-dependent protein kinase and CaM kinase II. One dimensional phosphopeptide mapping cleaved by *Staphylococcus aureus* V8 protease was very similar to that of synapsin I. Synapsin I is a neuron-specific phosphoprotein which has demonstrated a property consistent with its proposed role in linking synaptic vesicles to the cytoskeleton and its involvement in the regulation of neurotransmitter release (8). Our results showed that ACAMP-81 also plays an important role in neurotransmitter release, although further study is required to understand its physiological function. Another property of ACAMP-81 is that the

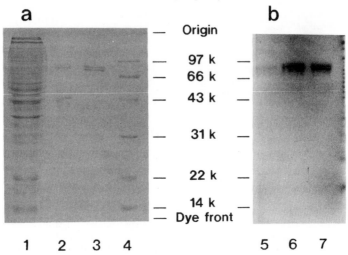

Fig. 3.   ACAMP-81 binding protein, protein staining (a), and ^{125}I-ACAMP-81 overlay (b). 0.6 M NaCl extract of bovine brain (lanes 1 and 5), partially purified ACAMP-81 binding protein (lanes 2 and 6), purified synapsin I (lanes 3 and 7), and molecular weight markers (lane 4) were electrophoresed on 12.5% SDS-PAGE, and processed for ^{125}I-ACAMP-81 overlay.

protein is also a good substrate for various protein kinases. Protein kinase C incorporated 1.46 mol phosphate into mol of ACAMP-81. Both cAMP-dependent protein kinase and CaM kinase II phosphorylated ACAMP-81 (0.75 mol P_i/mol protein, 0.46 mol P_i/mol protein, respectively).

We previously have synthesized the inhibitor of CaM kinase II (KN-62) and tested the effect of KN-62 on ACAMP-81 phosphorylation by CaM kinase II, finding that KN-62 inhibited the phosphorylation dose dependently.

II. CaM KINASE II INHIBITOR, KN-62 (9)

KN-62 (1-[N,O-bis-5-isoquinolinesulfonyl]-N-methyl-L-tyrosil-4-phenylpiperazine) is an isoquinoline sulfonamide derivative with a specificity against various protein kinases as described in Table I. The activities of protein kinase C, cAMP-dependent protein kinase and myosin light chain kinase were not inhibited by KN-62, however, it did specifically inhibit the activity of CaM kinase II with 0.3 μM of IC_{50} value.

One interesting property of CaM kinase II is autophosphorylation, which converts it into a calmodulin independent form. We next measured the effect of KN-62 on autophosphorylation of the enzyme and found that it inhibited the autophosphorylation of both α and β subunits of CaM kinase II dose dependently as did substrate phosphorylation. KN-62 did not inhibit the calmodulin independent activity of CaM kinase II after autophosphorylation, however, suggesting that KN-62 is a specific inhibitor of CaM kinase II and interferes with interaction of CaM kinase II with calmodulin.

Kinetic analysis of the inhibitory effect of KN-62 revealed as expected that it inhibits the activity of CaM kinase II competitively with respect to calmodulin, but not to ATP. To test whether KN-62 is a direct inhibitor or a calmodulin antagonist, we prepared the substance coupled in a Sepharose 4B column (Fig. 4). When a mixture of CaM kinase II and

TABLE I.
Specificity of KN-62

Kinase	IC_{50} (μM)
Ca²⁺/CaM dependent protein kinase II	0.3
Ca²⁺/phospholipid dependent protein kinase (PKC)	>100
cAMP dependent protein kinase	>100
Myosin light chain kinase	>100

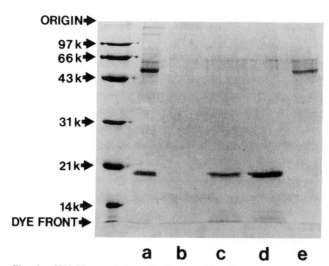

Fig. 4. KN-62 coupled to Sepharose 4B affinity chromatography. Rat brain Ca^{2+}/CaM kinase II (73 μg) and calmodulin (12 μg) in a solution (300 μl) containing 0.2 M NaCl, 40 mM Tris-HCl (pH 7.5), and 2 mM EGTA was applied to KN-62 coupled to Sepharose 4B (500 μl) which had been equilibrated with the above solution, then washed with 300 μl of this solution containing 1 M NaCl. Elution was carried out by adding 300 μl of SDS-PAGE sample buffer containing 8 M urea and boiling for 2 min. 40 μl of each fraction was applied to 12.5% SDS-PAGE. Lane a, applied sample; lanes b and c, void fractions; lane d, washed fraction with 1 M NaCl, 40 mM Tris-HCl (pH 7.5), 2 mM EGTA; lane e, eluted fraction. Left lane, molecular size markers.

calmodulin was applied to the column, only calmodulin went through without binding (lanes b, c, d). CaM kinase II could be eluated from the column by boiling with urea containing buffer (lane e). These results clearly indicated that KN-62 is a direct inhibitor of CaM kinase II and binds to the calmodulin binding site of the enzyme.

Interestingly, myosin light chain kinase is also a calmodulin stimulated enzyme, but was not inhibited by KN-62. This offers the possibility that KN-62 recognized the calmodulin-binding site of only CaM kinase II. We tested the effect of KN-62 on CaM kinase II activity in intact cells, and also measured the autophosphorylation of the enzyme which was immunoprecipitated with anti-CaM kinase II antibody from ^{32}P-labeled PC12D cells (pheochromocytoma cells). Ca-ionophore stimulated the phosphorylation of 53 kDa precipitated protein in the absence of KN-62. The 53 kDa phosphoprotein seemed to be the result of phosphorylation of CaM kinase II in PC12D cells. In the presence of KN-62, phosphorylation of the 53 kDa protein stimulated by Ca-ionophore was reduced dose dependently by addition of KN-62 to the culture medium.

We confirmed that KN-62 inhibits the activity of CaM kinase II not only *in vitro* but also in intact cells. CaM kinase II is thought to be widely distributed in various tissues and involved in various Ca²⁺-dependent cellular responses. KN-62 is thus seen as a useful tool to elucidate the physiological role of CaM kinase II in various tissues and cells.

SUMMARY

ACAMP-81 was purified from Triton X-100 soluble fraction of bovine brain and its molecular weight was estimated to be 81,000, and pI value 4.3. Alkaline gel electrophoresis revealed that ACAMP-81 bound to an equimolar amount of calmodulin, S-100 protein and troponin C in a Ca²⁺ dependent manner. Furthermore, an ACAMP-81 associated protein was detected and partially purified from bovine brain, and was identified as synapsin I. Since synapsin I is involved in neurotransmitter release, ACAMP-81 also plays an important role in this action. It is also a good substrate for various protein kinases. KN-62, newly synthesized CaM kinase II inhibitor inhibited ACAMP-81 phosphorylation, and its inhibitory mechanism was discussed.

REFERENCES

1. Kakiuchi, S., Yamazaki, R., and Nakajima, H. *Proc. Jpn. Acad.*, **46**, 587 (1970).
2. Brostorm, C.O., Huang, Y.C., Breckenridge, B.M., and Wolff, D.J. *Proc. Natl. Acad. Sci. U.S.A.*, **72**, 64 (1975).
3. Dabrowska, R., Sherry, J.M.F., Aromatorio, D.K., and Hartshorne D.J. *Biochemistry*, **17**, 253 (1978).
4. Yamauchi, T. and Fujisawa, H. *FEBS Lett.*, **116**, 141 (1980).
5. Kotani, S., Nishida, E., Kumagai, H., and Sakai, H. *J. Biol. Chem.*, **260**, 10779 (1985).
6. Tokumitsu, H., Mizutani, A., Nomura, S., Watanabe, M., and Hidaka, H. *Biochem. Biophys. Res. Commun.*, **163**, 581 (1989).
7. Mizutani, A., Tokumitsu, H., and Hidaka, H. *Biochem. Biophys. Res. Commun.*, **182**, 1395–1401 (1992).
8. Decamili, P. and Greengard, P. *Biochem. Pharmacol.*, **35**, 4349 (1986).
9. Tokumitsu, H., Chijiwa, T., Hagiwara, M., Mizutani, A., and Hidaka, H. *J. Biol. Chem.* **265**, 4315 (1990).

Regulation of Blood Cell Differentiation

TATSUO KINASHI, KWANG HO LEE, KAORU TOHYAMA, KEI TASHIRO, AND TASUKU HONJO

Department of Medical Chemistry, Kyoto University Faculty of Medicine, Kyoto 606, Japan

The totipotent bone-marrow stem cell is defined as a cell that has the potential to reproduce itself as well as to give rise to cells of each lineage in the blood. During the course of differentiation, progeny of the totipotent stem cell gradually lose their capacity to differentiate into particular lineages, giving rise to multipotent stem cells which eventually become committed to a single cell lineage. Stem cells proliferate, differentiate, and mature in the presence of growth factors. The totipotent bone-marrow stem cell has been difficult to maintain *in vitro* because a growth factor for this stem cell has not been isolated. On the other hand, molecular cloning of cDNAs encoding growth factors and lymphokines and their large-scale production have facilitated establishment of multipotent stem cell and progenitor cell lines cultured *in vitro* (reviewed in refs. *1* and *2*). Interleukin 3 (IL-3) has been shown to support the growth of multipotent stem cells without induction of differentiation (*3, 4*). There are several progenitor cell lines such as FDCP-mix A4 and 32DC13 which can be induced to a specific lineage by a lineage-specific growth factor, although most of them proliferate and remain undifferentiated in media containing IL-3 (*5, 6*).

Another regulatory element of hematopoiesis could be a stromal cell network in the bone marrow. This idea is supported by the finding that bone-marrow stromal cells are a necessary requirement for a long-term bone-marrow culture system, in which hemopoiesis of various cell lineages can be sustained without added growth factors (*1, 7*). Interestingly, the hemopoietic cells must be in direct contact with the stromal cells for proliferation and differentiation (8). Recently, the stromal cell lines PA6 and ST2 (9–11) were shown to support differentiation of primary bone-marrow cells into distinct lineage cells (reviewed in ref. 12). Such functional diversity of stromal cells could be ascribed to different growth factors produced or different cell adhesion molecules expressed by each stromal cell. Combination of appropriate growth factors derived from stromal cells and growth factor receptors expressed by progenitor cells might be a key element for the lineage-specific differentiation. In addition, an appropriate pair of cell adhesion molecules expressed on stromal and stem cells could be required for not only proliferation signal triggering but also specific growth factor/receptor expression.

It is important to know whether the lineage-specific differentiation induction by a stromal cell line is due to selection of progenitors which are already committed. Alternatively, the stromal cell line may direct differentiation lineages of a multipotential stem cell. To address this question we took advantage of an IL-3 dependent stem cell clone, LyD9 that has been shown to differentiate into multiple lineage cells, *i.e.*, B lymphocytes, neutrophils and macrophages by coculture with bone-marrow stromal cells (13–15). Differentiation of LyD9 cells requires direct contact with primary stromal cells (14). We have found that two stromal cell lines PA6 (9) and ST2 (11) are able to induce differentiation of LyD9 cells predominantly into different lineage cells, *i.e.*, granulocyte-macrophage colony stimulating factor (GM-CSF)-responsive and granulocyte (G)-CSF-responsive cells, respectively. The results suggest that differentiation lineages induced from a single stem cell clone are determined by cocultured stromal cell lines.

We then introduced the human M-CSF receptor gene into the LyD9, L-GM and L-G3 lines, and tested whether M-CSF could direct differentiation of c-*fms* transfectants of LyD9 cells and its derivatives into macrophages. We found that the M-CSF receptor expression alone did not allow these cells to differentiate into macrophages in response to M-CSF. We also found that premature expression of the M-CSF receptor on LyD9 and L-GM cells could not affect the differentiation lineage that is directed by the specific stromal cell lines (PA6 and ST2) used for the coculture. These results indicate that expression of a lineage-specific growth factor receptor alone is not sufficient for lineage commitment of the multipotential cell and that coculture with specific stromal cells is more deterministic to differentiation lineages.

I. GENERATION OF MYELOID CELLS FROM LyD9 CELLS BY COCULTURE WITH STROMAL CELL LINES

Differentiation of LyD9 cells by coculture with the stromal cell line PA6 was investigated in morphological and cytochemical studies. A considerable proportion (about 0.1%) of LyD9 cells cocultured with PA6 cells for 4 weeks contained myeloid-like nuclei and became positive for myeloid-specific enzymes like α-naphthyl esterase and myeloperoxidase. Similar experiments using ST2 stromal cells did not yield a detectable number of myeloid cells.

In order to characterize differentiated cells, an aliquot (10^6 cells) of LyD9 cells cocultured with the stromal cell line (either PA6 or ST2) was transferred every week to medium containing either GM-CSF or M-CSF. Viable cell numbers were scored 10 days after the transfer to growth factor-containing media. As shown in Table I, 4-week coculture of LyD9 cells with PA6 cells generated a good number of GM-CSF-responding cells and a small number of G-CSF-responding cells. By contrast, when PA6 cells were replaced by ST2 cells, no GM-CSF-responding cells were generated whereas a small number of G-CSF-responding cells were yielded (Table I). Coculture of LyD9 cells with the PA6 and ST2 stromal lines did not generate either B-lymphocytes or M-CSF-responsive cells even when various growth factors such as IL-4, IL-5, IL-7, and M-CSF were supplemented. Takeda et al. (15) reported that LyD9 cells can be induced to B lymphocytes by coculture with the RPO.10 stromal cell. The

TABLE I

Generation of GM-CSF and G-CSF-responding Cells by Coculture of LyD9 Cells with Stromal Cell Lines

Pretreatment of LyD9 cells	Weeks of coculture	PA6		ST2	
		+GM-CSF	+G-CSF	+GM-CSF	+G-CSF
None	2	0	0	0	0
	3	<100	<100	0	<100
	4	6×10^5	700	0	500
5-Azacytidine	2	1×10^3	300	0	200
	3	1×10^6	4×10^3	0	8×10^3
	4	5×10^7	6×10^4	500	2.5×10^5

LyD9 cells with or without 5-azacytidine pretreatment were cocultured with either ST2 or PA6 stromal cell line for weeks indicated, and 1×10^6 LyD9 cells were transferred to the medium containing GM-CSF (20 units/ml) or G-CSF (100 units/ml). Ten days after the transfer, viable cells were counted. Neither the experiment with M-CSF nor the same experiment without addition of the growth factor gave rise to any surviving cells (data not shown). Aliquots taken after one-week coculture did not yield any viable cells.

results support the hypothesis that differentiation lineage of the multi-potential cell line is determined by the cocultured stromal cell.

II. ESTABLISHMENT AND DIFFERENTIATION OF GM-CSF- AND G-CSF-DEPENDENT CELL LINES FROM LyD9 CELLS

LyD9 cells which had been cocultured with PA6 stromal cells were cultured in the medium containing GM-CSF for more than 4 weeks, and thereafter grown cells were maintained as a cell line called L-GM. Similarly, the L-G line was generated from 5-azacytidine-treated LyD9 cells by coculture with the ST2 stromal line, and subsequent selection in the G-CSF-containing medium. L-GM and L-G cells grew in the presence of GM-CSF and G-CSF, respectively; however, the growth response of L-G cells to G-CSF was transient. As L-GM and L-G cells proliferated in response to IL-3 like all the other derivatives of LyD9 cells (16), they were maintained in IL-3-containing media. No other growth factors examined could induce proliferation of L-GM and L-G cells.

Cytochemical studies were done to characterize differentiation properties of the L-GM and L-G lines cultured in GM-CSF and G-CSF, respectively (Fig. 1). About one third of the L-GM cells, which had been cultured in the presence of GM-CSF, had bent nuclei with a small nuclear-cytoplasmic ratio (Fig. 1a). L-GM cells were positive for α-naphthylbutyrate esterase and myeloperoxidase (Fig. 1b and c). A few neutrophils (less than 0.01%) were found in L-GM cells. L-GM cells showed the phagocytic activity when stimulated with phorbol 12-myristate 13-acetate (PMA) (Fig. 1d). A few percent of the L-GM cells appeared as mature macrophages even when cultured in IL-3 for one month. By contrast, the majority of the L-G cells, which had been cultured for 10 days in the presence of G-CSF, became neutrophils with typical lobulated nuclei and naphthol AS-D chloroacetate esterase-positive granules (Fig. 1e and f).

When L-GM cells were further cocultured with ST2 cells in the presence of 25 μg/ml lipopolysaccharide (LPS) for one month, about 30% of the cocultured L-GM cells had the morphological and cytochemical properties of neutrophils. Similar experiments with another GM-CSF-dependent derivative of LyD9, K-GM (16) yielded the same results (data not shown). Since monocyte/macrophage lineage cells appeared in L-GM and K-GM cells which had been cultured with GM-CSF, these results indicate that both L-GM and K-GM cell lines are progenitors of neutrophils and macrophages.

L-G cells proliferated and differentiated into neutrophils in the

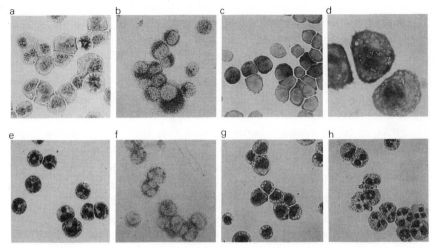

Fig. 1. Morphological and cytochemical analyses of L-GM and L-G cell lines. May-Grünwald-Giemsa staining of L-GM (a) and L-G (e) cell lines is shown. L-GM cell line (b) and L-G cell line (f) were analyzed cytochemically with both α-naphthylbutyrate esterase and naphthol AS-D chloroacetate esterase. Expression of myeloperoxidase (c) and functional phagocytosis (d) of latex beads in L-GM cell line was examined. Morphology of the L-G cell line maintained with IL-3 (5% of WEHI 3B conditioned medium) for one month is shown in g. The IL-3-cultured L-G cells were transferred to the medium containing G-CSF (100 units/ml), cultured for 10 days and their morphology examined (h).

presence of G-CSF, but gradually ceased to proliferate and the majority of them died in 2 weeks. L-G cells grew almost indefinitely when cultured in the medium containing IL-3. However, the morphological characteristics of neutrophils disappeared in the IL-3-containing medium in one month (Fig. 1g). The neutrophil phenotypes like lobular nuclei were regained by transferring L-G cells to the G-CSF-containing medium (Fig. 1h). The recovery of L-G cells was not due to reversal of maturation but rather to growth of remaining immature L-G cells, because careful limiting dilution experiments indicated that mature granulocyte clones derived from L-G cells by G-CSF did not proliferate in response to IL-3. It is likely that not all the L-G cells matured within 2 weeks in the G-CSF-containing media, and surviving immature L-G cells proliferated again in the presence of IL-3.

III. DIRECT CONTACT WITH STROMAL CELLS IS REQUIRED FOR MYELO-POIESIS FROM LyD9 CELLS

We then tested whether or not the direct contact of LyD9 cells to the stromal cell lines is also required for myelopoiesis as previously described for lymphopoiesis from LyD9 cells by coculture with primary bone-marrow stromal cells (*14*). We cultured LyD9 cells and the stromal cell line (PA6 or ST2) in two chambers separated by a nitrocellulose membrane (0.45 μm). Every week during this coculture, an aliquot of LyD9 cells (1×10^6 cells) was transferred to medium containing either GM-CSF or G-CSF, and grown cell numbers were counted 10 days later. As shown in Table II, no GM-CSF-responding cells and very few, if any, G-CSF-responding cells appeared even after 4 weeks of the coculture of LyD9 cells with the stromal cell lines without direct contact. The results clearly indicate that direct contact with stromal cells is important for differentiation of LyD9 cells into myeloid cells.

IV. INTRODUCTION AND EXPRESSION OF HUMAN C-*FMS* GENE IN LyD9 AND ITS DERIVATIVES

To test whether expression of the M-CSF receptor would modify differentiation of LyD9, L-GM and L-G3 cells, we introduced the human c-*fms* gene into these cells. We used the wild-type (Y)[996] and mutant-type (F)[996] of the c-*fms* gene (*17*). This replacement of tyrosine with phenylalanine at residue 996 augments M-CSF-dependent response without transformation. Transfected cells obtained were selected with G418 in medium containing IL-3. Two weeks after transfection, G418 resistant cells obtained were stained with monoclonal anti-human c-*fms* antibody and subjected to single cell sorting for cloning. Surface expression of the M-CSF receptor on cloned transfectants was confirmed by FACS analy-

TABLE II

Generation of GM-CSF and G-CSF-responding Cells by the Culture of LyD9 Cells with Stromal Cell Lines in Separate Membrane Chambers

Pretreatment of LyD9 cells	Weeks of coculture	PA6		ST2	
		+GM-CSF	+G-CSF	+GM-CSF	+G-CSF
None	2	0	0	0	0
	3	0	<100	0	<100
	4	0	<100	0	<100

Experiments were done as described in Table I except that LyD9 and stromal cells were cultured in separate membrane chambers (pore size 0.45 μm).

sis. To see whether the expressed c-*fms* proteins were functional, we performed an *in vitro* kinase assay. Two phosphorylated proteins were generated in all transfectants.

V. GROWTH AND DIFFERENTIATION OF C-*FMS* TRANSFECTANTS BY M-CSF

We then investigated growth responsiveness of the c-*fms* transfectant to human M-CSF. All three LyD9/*fms* clones failed to grow. They survived slightly longer than LyD9 cells, but all cells were dead by day 3 (Fig. 2A). LyD9/*fms*(F) clones showed a transient growth surge for 2 days and declined rapidly. LyD9/*fms*(F) cells showed slightly augmented responses to M-CSF as expected from the observation that the replacement of tyrosine at residue 996 with phenylalanine increases the size and the number of M-CSF colonies of NIH3T3 bearing c-*fms* (*17*). While LS-1 cells displayed M-CSF-dependent growth, it is clear that the M-CSF receptor expressed on LyD9 cells failed to convey signals for continued proliferation.

The same pattern was observed for all but one of the L-GM*fms*/ clones (Fig. 2B). L-GM/*fms*(F) showed a transient response pattern, but

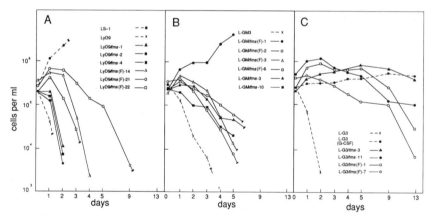

Fig. 2.   Growth responses to M-CSF. Growth profiles of c-*fms* transfectant of clones LyD9(A), L-GM(B), and L-G3(C) cells in the presence of M-CSF (50 units/ml). 1–2 × 10 cells were grown in 5 ml of complete medium with or without M-CSF and the medium was changed to a fresh one every third day. Viable cells were counted with the trypan blue exclusion. LS-1 cells were included as a positive control and untransfected parent cells as a negative control. Growth patterns of c-*fms*-transfected clones without M-CSF were similar to those of parent cell lines. In C, growth response of L-G3 cells to G-CSF is included for comparison.

declined thereafter. L-GM/*fms* showed a similar response to a lesser degree. One clone of L-GM/*fms*(F)-1 continued to proliferate in the presence of M-CSF. It expressed a similar amount of c-*fms* proteins as other L-GM/*fms* clones, and its growth is still dependent on M-CSF. We have not characterized any further difference between L-GM/*fms*(F)-1 and the other clones. As the L-GM/*fms*(F)-1 phenotype is relatively rare, secondary events might be required for c-*fms* transfectants of LyD9 and L-GM cells to acquire the proliferation capability in response to M-CSF.

In contrast, all L-G3/*fms* clones showed prolonged survival in the presence of M-CSF (Fig. 2C). L-G3 transfectants bearing either wild or mutant type M-CSF receptors doubled in number after 24 hr and maintained their viability for up to 2 weeks. G-CSF induced a similar growth profile except for a longer period up to 18 days.

We tested whether M-CSF could alter morphological and cyto-chemical properties of the c-*fms* transfectants. LyD9/*fms* cells showed immature morphology in the presence of M-CSF and did not change this phenotype on day 2, one day before death (data not shown). L-GM/*fms* cells cultured in the presence of either IL-3 or GM-CSF contained a small fraction of neutrophils and macrophages among immature cells. Culturing in M-CSF did not change this morphology of the L-GM/*fms* cells on day 3-4. Most L-GM/*fms*(F)-1 cells, which proliferated continuously in M-CSF, were immature with few mature macrophages and neutrophils (data not shown).

When L-G3/*fms* cells were cultured with G-CSF, these cells began to differentiate about day 7, and most of them became typical mature neutrophils by day 14 (Fig. 3c and h). When L-G3/*fms* cells were cultured in M-CSF-containing media, they showed not only similar growth responses (Fig. 2c), but also similar differentiation profiles (Fig. 3d and i) to those induced by G-CSF. Morphological changes became apparent around day 7. Mature neutrophils accounted for 20-50% of the population between day 9 and day 14. The remaining cells were still immature without monocytic morphological properties such as enlarged and vacuolated cytoplasm, non-specific esterase, and phagocytic activities, which were typically found in LS-1 cells treated with M-CSF. L-G3 transfectants bearing the mutant form(F) of c-*fms* tended to produce more neutrophils than those bearing the wild form(Y) (Fig. 3e and j). L-G3/*fms* remained immature in IL-3.

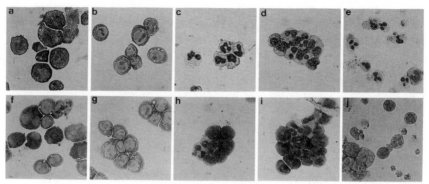

Fig. 3.   Histological examination of c-*fms* transfected clones. May-Grünwald-Giemsa stain from a to e and myeloperoxidase staining from f to j. a and f, LyD9/*fms*-1/GM in GM-CSF (10 units/ml); b and g, L-G3/*fms*-3 in IL-3 (50 units/ml); c and h, L-G3/*fms*-3 in G-CSF (50 units/ml); d and i, L-G3/*fms*-3 in M-CSF (50 units/ml); e and j, L-G3/*fms*-(F)-1 in M-CSF. For the culture with G-CSF and M-CSF, cells were harvested and stained on day 10. L-G3/*fms*-(F)-1 in IL-3 and G-CSF showed similar morphology to those of L-G3/*fms*-3.

VI.   DIFFERENTIATION OF LyD9/*FMS* ON STROMAL CELLS

Finally, we tested whether earlier expression of the M-CSF receptor on LyD9 cells would affect the differentiation lineage determined by coculture with stromal cells. LyD9/*fms*(Y)-1 cells were cultured with PA6 stromal cells in the presence of a minimal amount of IL-3 for 2 weeks. Nonadherent cells were harvested and maintained with GM-CSF. Two clones were established by limiting dilution. LyD9/*fms*-1/GM and LyD9/*fms*-2/GM clones were dependent on either GM-CSF or IL-3. Most of them proliferated without differentiation in the presence of IL-3 or GM-CSF, but some of the cells did differentiate into macrophages and neutrophils (Fig. 3a and f). Neither LyD9/*fms*-1/GM nor LyD9/*fms*-2/GM cells was induced to proliferate and differentiate in response to human M-CSF. No cell lines were obtained even when nonadherent cells were cultured in human M-CSF.

LyD9/*fms*-1 cells were cocultured with ST2 stromal cells to induce G-CSF-dependent lines. After 2 weeks of coculture, nonadherent cells were harvested and maintained with G-CSF for 2 days, and then expanded and maintained with IL-3. LyD9/*fms*-1/G responded to both G-CSF and M-CSF with the same prolonged survival pattern as found in L-G3/*fms* cells. LyD9/*fms*-1/G cells survived for about 18 days in G-CSF but for about 12 days in M-CSF. Morphological studies showed that most LyD9/*fms*-1/G cells differentiated into neutrophils in the presence of G-CSF, whereas 30–40% of the cells were neutrophils in the presence of M-CSF. Identification of neutrophils was based on ringed or segmented nuclei, and myeloperoxidase production. Similar lines were obtained when M-CSF was added in place of G-CSF for 2 days before maintaining the cells in IL-3.

These results indicate that premature expression of the M-CSF receptor does not direct the monocytic commitment of LyD9, L-GM, and LG3 clones, nor does it alter differentiation lineage determined by stromal cells cocultured with the multipotential stem cell.

SUMMARY

Coculture with the cloned stromal cell lines PA6 and ST2 can induce differentiation of an IL-3 dependent multipotential stem cell clone, LyD9 cells, predominantly into GM-CSF- and G-CSF-responding cells, respectively. The stromal cell-induced differentiation of LyD9 cells required direct contact between LyD9 and stromal cells. These results suggest that the differentiation lineages of the LyD9 stem cell are determined by the stromal lines. To test whether expression of a lineage-specific growth factor receptor is deterministic to lineage commitment during hematopoiesis we introduced the human c-*fms* gene into LyD9 and two myeloid progenitor lines derived from LyD9 cells. We showed that differentiation lineage is not affected by premature expression of the M-CSF receptor. Instead, the stromal cells used for coculture with multipotential stem cells apparently control determination of lineage commitment.

REFERENCES

1. Dexter, T.M. and Spooncer, E. *Annu. Rev. Biol.*, **3**, 423 (1987).
2. Nicola, N.A. *Annu. Rev. Biochem.*, **58**, 45 (1989).
3. Metcalf, D., Johnson, G.R., and Burgess, A.W. *Blood*, **55**, 138 (1989).
4. Ihle, J.N., Rebar, L., Keller, J., Lee, J.C., and Hapel, A. *J. Immunol. Rev.*, **63**, 5 (1982).
5. Heyworth, C.M., Dexter, T.M., Kan, O., and Whetton, A.D. *Growth Factors*, **2**, 197 (1990).
6. Valtieri, M., Tweardy, D.J., Caracciola, D., Johnson, K., Malvilio, F., Altmann, S., Santoli, D., and Rovera, G. *J. Immunol.*, **138**, 3829 (1987).
7. Whitlock, C.A. and Witte, O.N. *Proc. Natl. Acad. Sci. U.S.A.*, **79**, 3608 (1982).
8. Bentley, S.A. *Exp. Hematol.*, **9**, 308 (1981).

9. Kodama, H.-A., Amagai, Y., Yamada, H., and Kasai, S. *J. Cell Physiol.*, **112**, 83 (1982).
10. Nishikawa, S.I., Ogawa, M., Nishikawa, S., Kunisada, T., and Kodama, H. *Eur. J. Immunol.*, **18**, 1767 (1988).
11. Ogawa, M., Nishikawa, S., Ikuta, K., Yamamura, F., Naito, M., Takahashi, K., and Nishikawa, S.-I. *EMBO J.*, **7**, 1337 (1988).
12. Kincade, P.W., Lee, G., Pietrangeli, C.E., Hayashi, S.I., and Gimble, J.M. *Annu. Rev. Biochem.*, **7**, 111 (1989).
13. Palacios, R. and Steinmetz, M. *Cell*, **41**, 376 (1988).
14. Kinashi, T., Inaba, K., Tsubata, T., Tashiro, K., Palacios, R., and Honjo, T. *Proc. Natl. Acad. Sci. U.S.A.*, **85**, 4473 (1988).
15. Takeda, S., Gillis, S., and Palacios, R. *Proc. Natl. Acad. Sci. U.S.A.*, **86**, 1634 (1989).
16. Kinashi, T., Tashiro, K., Inaba, K., Takeda, T., Palacios, R., and Honjo, T. *Int. Immunol.*, **1**, 11 (1989).
17. Roussel, M.F., Dull, T.J., Rettenmier, C.W., Ralph, P., Ullrich, A., and Sherr, C.J. *Nature*, **325**, 549 (1987).

Probe Studies of the Orientation of the Myosin Cross-Bridge during Contraction

THOMAS P. BURGHARDT AND KATALIN AJTAI

Department of Biochemistry and Molecular Biology, Mayo Foundation, Rochester, Minnesota 55905, U.S.A.

The generation of force to produce muscle shortening against a load takes place during the hydrolysis of the nucleotide ATP inside a muscle. The mechanism of muscle contraction wherein the chemical energy from ATP hydrolysis is converted into mechanical work is contained within the structures of the myosin and actin. Since the site of the ATPase is on the myosin head or cross-bridge, the energy transduction must involve structural changes within the cross-bridge. The nature of these structural changes and how they make muscle work is one of the central questions of muscle research.

It is well known that the occupancy of the ATP binding site modulates cross-bridge affinity for actin. When ATP is absent the cross-bridge binds tightly to actin producing a rigor muscle that is mechanically very stiff. When ATP binds to the cross-bridge the equilibrium shifts to the dissociated state and the muscle relaxes. Muscle contraction occurs when Ca^{2+} and ATP are present. These observations suggest that the myosin and actin interact cyclically during contraction and that the cross-bridges may act as independent generators of force (*1*).

A popular and useful model for the role of myosin and actin in contraction is the rotating cross-bridge model (*2, 3*), wherein the myosin head rotates as a rigid body while attached to the actin filament during ATP hydrolysis. This mechanism can be accommodated by a cross-bridge that contains two or more spatially separated actin binding sites with the actin affinities of the separated sites modulated (so as to produce a rotation) by the state of the nucleotide in the nucleotide binding site. We will show subsequently that much of the physical data related to cross-bridge orientation is interpretable in terms of this basic rotating cross-bridge model.

We also want to investigate how adequately a rigid body rotation of the cross-bridge describes the cross-bridge participation in contraction. It seems certain that the cross-bridge undergoes complex structural changes during contraction that are only approximately described as a rigid body rotation. The structural changes within the head in response to ATP hydrolysis could be part of the transmission of energy from the ATPase site to the actin binding site and not involving rotation, or relevant to the direct production of force at the actomyosin interface (as in relative rotation of domains within the myosin head). The techniques we employ to quantitate cross-bridge dynamics can investigate the global movements (rigid body-like), and the intramolecular movements of domains of the myosin cross-bridge.

In the following pages we introduce the basic theory of the physical techniques for detecting probe orientation from chemically modified cross-bridges in muscle fibers. We discuss some of the biochemical problems in the specific modification of muscle proteins and a new method for verifying specific labeling. Finally, we discuss the applications of probe techniques, and point out the new directions for investigation of the molecular mechanism of muscle contraction.

I. OVERVIEW

We use two physical methods for detecting signals from extrinsic probes. One of them, electron paramagnetic resonance (EPR), detects the absorption of light by free radicals whose spins are oriented by a large magnetic field. The second, fluorescence polarization, detects the polarization and intensity of light emitted by a probe molecule after excitation by polarized light. Both techniques are understood in a fundamental manner (4). In sections II and III we give a brief explanation of the physical basis of the techniques.

We are also concerned with biochemical methods pertaining to the specific chemical modification of protein side chains. Getting reliable information from signals originating from probes depends on our ability to place the probes specifically in the region of interest. These methods involve the thoughtful choice of probes based on probe shape or charge or other probe characteristics and the ability to detect where a probe is located on the polypeptide chain after the protein has been modified. We will summarize the principles of these methods and describe an application to myosin in section IV.

In section V, we discuss the interdependence of the probe techniques for determining probe order. We want to merge the data from different techniques when they are used for detecting probe orientation. In our application this means combining data from spin and fluorescent probes of myosin. If we describe probe order as a linear combination of order parameters, merging data from different techniques is a standard problem of linear algebra. We will introduce this formalism and summarize what we can expect to find out from this powerful analytical technique about the mechanism of contraction.

II. EPR

Probes with a detectable EPR signal are generally molecules with an unpaired electron, *i.e.*, a radical species. For biochemical studies the most often used EPR signal donors are the relatively stable nitroxide probes. A variety of probes are commercially available to suit different applications (5). The unpaired electron in the nitroxide probe is a spin, $S = 1/2$, particle that interacts strongly with a magnetic field *via* the Zeeman interaction (4). In an EPR experiment a large static magnetic field, H (the Zeeman field), is applied to the sample containing the spin probes. The H-field splits the energy levels of the electron spins such that spins parallel to H are lower in energy than spins perpendicular to H and the energy separation depends linearly (to lowest order) on the magnitude of H. The transition from the lower to the higher energy orientation corresponds to the absorption of a quantum of light of appropriate frequency from a second (after the Zeeman field), oscillating, magnetic field. Typically the frequency of the oscillating field is constant and the value of the Zeeman field is slowly swept through the absorption resonances. The efficiency of absorption of energy from the oscillating field is related to the orientation and local environment of the spin probe (6).

III. FLUORESCENCE POLARIZATION SPECTROSCOPY

When a fluorescent molecule interacts with an incident external electromagnetic field the valence electron absorbs light energy with a probability proportional to the square of the absorption dipole, $\vec{\mu}_a = \langle S_j | \vec{x} | S_0 \rangle$, where $|S_j\rangle$ is the spatial wave function of excited state j, and $\vec{x}_a$ is the spatial coordinate of the valence electron. The excited state relaxes rapidly by radiationless transitions to the lowest energy excited state, S_1. The energy separation between S_1 and the ground state S_0 is large enough to delay further radiationless decay so that spontaneous fluorescence emission, with a natural lifetime of nanoseconds, contributes to the relaxation process. The probability of spontaneous relaxation to the ground state *via* fluorescence is proportional to the square of emission dipole, $\vec{\mu}_e = \langle S_1 | \vec{x} | S_0 \rangle$.

In a fluorescence polarization experiment probes are excited by linearly polarized light as a function of polarization and excitation wavelength. The emission is also detected as a function of wavelength and polarization. Absorption dipoles, $\vec{\mu}_a$, computed from different excited state wave functions $|S_j\rangle$ point in varying orientations relative to a probe fixed reference frame depending on excitation wavelength. Emission dipoles, $\vec{\mu}_e$, are computed from S_1 and S_0 and are generally less orientation dependent as a function of emission wavelength. The excitation polarization spectrum (measured by varying the excitation wavelength while keeping emission wavelength constant) is a measurement that takes advantage of the variable absorption dipole orientation (due to excitation wavelength variation) to detect possible changes in any of the orientational degrees of freedom available to the probe.

The steady-state fluorescence polarization experiment from an oriented biological assembly consists of the measurement of four intensities $F_{\|,\|}$, $F_{\|,\perp}$, $F_{\perp,\perp}$, and $F_{\perp,\|}$, where the first subscript refers to the excitation polarization and the second to emission polarization and $\|$ or $\perp$ means relative to a laboratory frame. In the muscle fiber studies we assume the lab frame to be fixed along the fiber axis such that $\|$ means parallel to the fiber axis. These intensities are combined into the three ratios $P_{\|} = (F_{\|,\|} - F_{\|,\perp})/(F_{\|,\|} + F_{\|,\perp})$, $P_{\perp} = (F_{\perp,\perp} - F_{\perp,\|})/(F_{\perp,\perp} \ F_{\perp,\|})$, $Q_{\|} = (F_{\|,\|} - F_{\perp,\|})/(F_{\|,\|} + F_{\perp,\|})$. We use ratios because they are independent of constants related to the intensity of the excitation light and the efficiency of the collection optics.

These ratios are related to the probe distribution by well understood, although complicated, mathematical formulas (7). Some features of the probe distribution can be surmised directly from the values of ratios. For instance, $P_{\perp} = 0$ indicates a random probe distribution, $P_{\perp} > 0$ indicates probes with transition dipoles more perpendicular than parallel to the fiber axis, and $P_\perp < 0$ indicates probes with transition dipole more parallel than perpendicular to the fiber axis. We distinguish different orientations of the muscle fiber cross-bridges with this simple test since it is rigorously correct to assert that a cross-bridge state transition that changes $P_\perp$ from >0 to <0 or *vice versa* is from cross-bridge rotation and not due to a disordering of the probe distribution. Distinguishing probe disordering from probe rotation is critical in the verification of the rotating cross-bridge model of contraction.

IV. BIOCHEMICAL TECHNIQUES FOR SPECIFIC LABELING

1. Covalent Modification of Proteins

It is essential to the probe studies of biological systems that the probe fit specifically and rigidly into a well defined point on the biomolecule of interest without altering its biological activity. In many cases including myosin, the amino acid side chains of the protein involved in biological activity have altered reactivities due to their irregular pK_a values. This gives you the possibility to introduce labels exclusively to these points. Lowering the pH of the reaction medium far below the pK_a of most of the lysine ε-amino groups makes these sites unfavorable for modification while cysteinyls, especially those having altered pK_a values, can still react. Sulfhydryl 1 (SH1) of myosin is a good example of a sulfhydryl exhibiting this phenomenon.

Evidently the three dimensional shape of the biomolecule can also influence probe specificity. The myosin cross-bridge and the other muscle proteins contain several free thiols but only SH1 is highly reactive in the absence of actin. We devised a protocol for the selective labeling of the second thiol (SH2) on cross-bridges in fibers using the sulfhydryl-selective label 4-[N-[(iodoacetoxy)ethyl]-N-methylamino]-7-nitrobenz-2-oxa-1,3-diazole (IANBD). The protocol promotes the specificity of IANBD by using the ability to protect SH1 from modification by binding the cross-bridge to the actin and using cross-bridge bound MgADP to promote the accessibility of SH2 (8).

The specific labeling of the complex muscle fiber system with spin probes follows the rules discussed for fluorescent labels. A special difficulty appears due to the poorer specificity of the traditionally used (maleimido)tempo label. An additional chemical treatment (potassium ferricyanide), that destroys unwanted spins not located at SH1, is needed to make the signal donor site specific (9).

2. Verifying Probe Specificity

Sodium dodecyl sulfate-polyacrylamide gel electrophoresis (SDS-PAGE) is a method of separating and identifying proteins and peptides by molecular weight (*10*). Fluorescent probes in the separated peptides are identified by their characteristic color since the probes are resistant to degradation in the SDS-PAGE. Consequently, with fluorescent probes the specificity of new labels is determined by optically detecting the fluorescent peptide bands on the gel and comparing the pattern to a Coomassie-stained digestion pattern.

Nitroxide radicals are degraded by the usual SDS-PAGE separation process so radioisotope forms of spin labels were routinely used to identify the probe in the separated peptides by following radio emissions. The isotope does not distinguish between the spin probe population with and without an electron spin. We developed a protocol to protect the intact spin label during the SDS-PAGE separation process (*11*). The spin labels are then detected in the gel by EPR. Our method allows the spin label specificity to be determined directly from the gel as it is done with fluorescent probes.

3. Biological Activity

The assessment of the damage caused by the incorporation of the covalent probe of myosin is a complicated issue. The enzymatic activity of the myosin cross-bridge is altered by the modification of SH1. It is well known that upon modification of SH1 the ATPase of myosin in the presence of Ca^{2+} is activated 4–8 fold while the ATPase in the absence of divalent ions, the so-called K^+-EDTA ATPase, is inhibited linearly with the fraction of SH1's modified (*12*).

In spite of the altered ATPases, the active isometric tension and the rigor tension of probe modified fibers is unchanged by modification of SH1 or SH2 (*13-15*). Researchers reasoned that the fiber tension shows that the contractile system remains essentially intact and that the altered ATPases reflect modifications in enzymatic steps of activated cross-bridges that are not rate limiting in muscle contraction. Unlike SH1 modification, the covalent modification of SH2 with a fluorescent probe does not alter the myosin ATPases (*12, 16*) or the isometric tension of active modified fibers (*8*). The orientation of SH2 changes upon changing physiological states of the fiber but not in a manner that parallels SH1 (*8, 17*). Clearly, modifying and detecting the orientation of other side chains on myosin will help to clarify the validity of measurements of SH1 orientation and are an important approach for testing the viability of the rotating cross-bridge model of muscle contraction.

V. THE INTERDEPENDENCE OF PROBE DATA

It occurred to us that the data concerning probe orientation distributions could be summarized in a general manner that is model-independent (7). For our purpose the probe angular distribution, N, is most conveniently described as a linear combination of order parameters. We showed that the fluorescence polarization signal depends on a limited number of order parameters and that the EPR signal depends on an infinite number of order parameters. This implies that EPR is a higher angular resolution technique than fluorescence polarization (18).

Any description of probe order assumes a particular molecular coordinate frame. Experimental data is most easily summarized in terms of a molecular frame that corresponds to a physically identifiable orientation intrinsic to the probe molecule. For fluorescent probes this "physical" frame is that which corresponds to the dipole moment of the excited probe, for spin labels it is the principal magnetic frame. These probe fixed frames are different for each probe but all molecular frames are related by a rotation of coordinates.

The angular relationship among probes of a given side chain (SH1 of myosin, for example) corresponds to a set of linear equations relating the probe order parameters to one another. We require that a rotation of the protein carrying the probes must rotate each of the separate probe coordinate frames equivalently. We can show that this requirement constrains the set of equations relating order parameters enough so that we can solve for the quantities relating to the probe molecular frames and for an even larger set of order parameters of each probe. This leads to a large improvement in the resolution of the probe angular distribution measurable from individual fluorescent and spin probes.

VI. EPR STUDIES OF MYOSIN CROSS-BRIDGES

EPR was used to study myosin cross-bridges many times. The earliest work used a sulfhydryl specific spin probe, (iodoacetamido) tempo spin labels (ITSL), to modify SH1 of myosin (19). In this work the sensitivity of the spin probe to its local environment was used to detect conformational changes in the vicinity of SH1 due to ATPase activity of the myosin in solution. SH1 was also the target of (maleimido)tempo spin labels (MTSL) that were used to study myosin cross-bridge orientation in muscle fibers.

Specifically labeling SH1 with MTSL is more difficult since the maleimido based probes are less specific for thiols than the iodoacetamide based probes (12). A ferricyanide wash, modified for use on fibers, destroyed spins not bound to SH1 (20). Spin labeled cross-bridges in ferricyanide treated muscle fibers appeared to be specifically labeled at SH1. The EPR spectra showed that cross-bridges in rigor were well oriented as expected from earlier fluorescent probe studies. Unlike previous work, however, fibers treated in this way failed to exhibit any cross-bridge order in relaxation (20). These spin labeling studies also failed to detect cross-bridge rotation upon binding MgADP. The most interesting result of this early work was that reported on contracting muscle fibers. The EPR spectrum from contracting fibers was reported to show that 80% of the cross-bridges were relaxed and probably detached from actin and that the remaining 20% were actin attached in an orientation identical to that present in rigor muscle (21).

More recent results derived from labeled muscle fibers do not confirm these early findings using EPR. We used MTSL labeled myosin subfragment 1 (MTSL-S1) to decorate muscle fibers in rigor and in the presence of MgADP. MTSL-S1 will bind tightly to muscle fibers in these conditions. Fibers decorated with MTSL-S1 show a high degree of order in rigor and in the presence of MgADP, however, the cross-bridge orientation in these two states differs significantly implying the cross-bridge rotates on actin upon binding nucleotide (22).

The decoration experiments eliminated the need to wash the entire fiber in ferricyanide. We found that the ferricyanide wash, done under the conditions of the original spin labeling protocol published for muscle fibers (20, 21), removes the Ca^{2+} regulation of muscle contraction. A less denaturing treatment to remove nonspecific spin label must be devised.

An alternative approach is to use a more specific label. The ITSL is more specific for SH1 but is unsuitable for studying cross-bridge orientation when the cross-bridge binds nucleotide since the nucleotide causes the probe to become mobile (23). We tried modifying SH1 with (iodoacetamido)proxyl spin label (IPSL) and found that the probe was very specific for SH1 (as is ITSL) and immobilized on the surface of myosin in the presence and absence of nucleotides (11). Studies of the orientation of IPSL-S1 decorating muscle fibers in rigor and in the presence of MgADP also show that the IPSL-S1 rotates from its orientation in rigor when the IPSL-S1 binds MgADP. This probe shows a large change in the shape of the spectrum when the cross-bridge binds MgADP. The spectra, from fibers with fiber axis parallel to the Zeeman field, indicate a 10–20 degree cross-bridge rotation in the rigor to MgADP state transition (*11*).

VII. STEADY-STATE FLUORESCENCE STUDIES OF MYOSIN CROSS-BRIDGES

Steady-state fluorescence polarization is a well established technique for monitoring cross-bridge orientation and conformation (*18*). Aronson and Morales studied the polarized emission from tryptophan residues in glycerinated muscle fibers (*24*). In these studies the polarized tryptophan emission following polarized excitation originated mainly from the myosin head. This work established that changes in the physiological state of the muscle fiber are detected as changes in the polarized emission from the tryptophans in S1.

There are many tryptophans in the proteins making up the glycerinated muscle fiber so the molecular origin of the changes observed in the tryptophan emission due to cross-bridge state changes were difficult to identify. Nihei *et al.*, used the fluorescent probe N-(iodoacetylamino-ethyl)-5-napthylamine-1-sulfonic acid (1,5-IAEDANS) to specifically label SH1 in S1 (*25*). This earliest work with a specific fluorescent probe of the myosin cross-bridge showed that the measured values of $P_\parallel$ and $P_\perp$ were distinct for rigor, relaxed and isometrically contracting fibers.

More recently other fluorescent and spin probes attached to SH1 were used to study cross-bridge orientation. Borejdo *et al.* reported that the SH1 bound probe (iodoacetamido)tetramethylrhodamine (IATR) maintained distinct orientations relative to the actin filament when the fiber is in rigor compared to that in the presence of MgADP (*26*). This observation was later confirmed and the study extended to active isometric fibers where it was found that active cross-bridges have an orientation distinct from that in either rigor or in the presence of MgADP (*27, 28*).

The results from the IATR labeled fibers were controversial since they seemed to conflict with other probe studies of SH1. The early spin labeled fiber studies (discussed in section I) indicated that the SH1's in activated isometric fibers are either randomly oriented (80%) or oriented as in rigor (20%) (*21*). The fluorescent probe 1,5-IAEDANS also modifies SH1 and was reported to not distinguish the rigor orientation from that in the presence of MgADP (*26*). One model that accommodates these results within a rotating cross-bridge model of muscle contraction is that the probes of SH1 have different orientations on the cross-bridge and that IATR points in a direction very favorable for detecting the cross-bridge rotation (*i.e.*, nearly perpendicular to the axis of rotation), while 1,5-IAEDANS and MTSL point in a very unfavorable direction for detecting cross-bridge rotation. A physical method for testing this model exists for the fluorescent probes. We measured $P_\perp$ as a function of excitation wavelength from 1,5-IAEDANS labeled S1 decorating fibers in rigor and in the presence of MgADP as shown in Fig. 1. If the decorated fiber is irradiated with polarized light of wavelength < 420 nm, the 1,5-IAED-ANS indicates a similar cross-bridge orientation in rigor and in the presence of MgADP (similar in the sense that $P_\perp < 0$ for both states). If the excitation wavelength is near 426 nm, $P_\perp > 0$ for cross-bridges in the presence of MgADP and $P_\perp < 0$ for rigor cross-bridges (*29*). These data indicate the cross-bridge rotates upon binding MgADP.

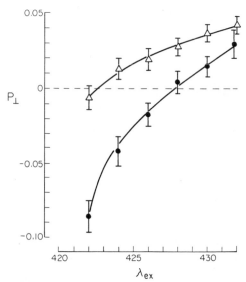

Fig. 1. $P_\perp$ measured from 1,5-IAEDANS labeled S1 decorating muscle fibers in rigor (●) and in the presence of ADP (△) as a function of excitation wavelength, λ_{ex}. λ_{ex} is measured in nanometers. Error bars indicate standard error of the mean. Reprinted with permission from American Chemical Society (From ref. *29*).

VIII. CONNECTING PROBE DATA SETS TO AMPLIFY ANGULAR RESOLU-
TION

The purpose of the probe studies we do on muscle is to deduce the angular distribution of the probe in a timely manner with the most angular resolution possible. We have remarked about the capabilities of the fluorescence polarization and EPR techniques for detecting probe order. Each technique supplies us with information that is incomplete and our job is to deduce what the data is telling us even though we don't have the total picture. With the limitations of the techniques in mind we realized that identifying what the theoretical limits are on what the probes can tell us about the probe angular distribution was our first objective. This theoretical limit turned out to be easy to define if you describe probe order in terms of the angular order parameters (7). We found that both the EPR and fluorescence polarization signals failed to detect certain well defined types of probe order and that, of the two techniques, EPR was able to give a more refined picture of probe order.

The fact that fluorescence polarization could be measured as a function of excitation wavelength meant that this technique had the unique property that a single fluorescent probe could be rotated at will within the molecular frame of the protein. It is clear, from the preceding discussion of the controversy surrounding probe data from SH1, how critical this property could be in removing the ambiguities from the probe measurements. It seems obvious that a variety of methods may be needed to get to the right picture of cross-bridge participation in muscle contraction and this brings us to a second objective. We would like to be able to formulate our description of the probe angular distribution in a general manner that is applicable to data from any technique. This we did by formulating the model-independent description of probe order using order parameters that are universally applicable to the probe techniques (7). A consequence of applying this approach to EPR and fluorescence polarization data is the ability to relate the data from these different techniques and thereby to increase the resolution of the angular distribution for all of the probes. This work is now in progress.

IX. CONCLUSIONS

The role of cross-bridge orientation in muscle contraction has been and continues to be a well investigated area of muscle research. The continued use of extrinsic direction reporting probes in this investigation is justified given that the biochemical techniques to introduce specific probes without harming biological activity are improving and that the biophysical techniques for detecting probe orientation, such as EPR and fluorescence polarization, are high resolution and have an intrinsically high signal-to-noise ratio.

Despite these advantages enjoyed by the extrinsic probe techniques, fluorescent and EPR probe studies of cross-bridge orientation produced contradictory interpretations of the data. We contend that the contradictions arose because of ambiguity in the information supplied by the probe data. No single probe can investigate all three angular degrees of freedom in which a myosin cross-bridge can move. The experimental technique or the precision with which the technique is used cannot compensate for the unfavorable alignment of the probe axis with the principal axis of rotation of the cross-bridge or the equivalence of a physical probe rotation with a rotational ambiguity of the signal detection method.

Our conclusions are that (i) the probe angular distribution from cross-bridges in rigor contains at least two distinct probe orientations, and (ii) that the rigor cross-bridge orientation can be perturbed into different actin bound orientations by the occupancy of the ATP binding site. We speculate that the myosin cross-bridge has two or more binding sites for actin and that each binding site causes the cross-bridge to have a unique orientation relative to actin. The active cross-bridge rotates by occupying the binding sites sequentially. The hydrolysis of ATP modulates the affinity of the various actin binding sites to drive the active cross-bridge rotation. The unambiguous result of the probe studies is that the cross-bridge can rotate when the fiber changes physiological states. This is a solid result and is consistent with all of the published data. It also brings us to the next level of sophistication of questions we wish to address with the probe orientation techniques. We want to know what the role is of the intra cross-bridge dynamics in muscle contraction. People have been studying the intramolecular dynamics of myosin and S1 for a long time using several techniques (*30*). This work has produced a great deal of information concerning the three dimensional structure of the myosin cross-bridge. The emphasis in this work is the study of the path and mechanism of energy transduction from the ATPase site to the actin binding site. The techniques used are generally thought to be sensitive to translational rather than orientational dynamics.

We propose to emphasize the orientational dynamics of the myosin cross-bridge intramolecular motions. The way to begin is to study the orientation changes of two or more points on the cross-bridge to see if their angular change is consistent with (i) a rigid body rotation of the entire myosin head, or (ii) a local deformation of the head. The comparison of data from SH1 and SH2 probes is a good starting point for this investigation. The analytical methods to join data from different probes and/or techniques offers a way to investigate whether or not the data are consistent with (i) or (ii) above.

SUMMARY

A working model for the molecular mechanism of muscle contraction has the myosin cross-bridge rotating while attached to actin to produce muscle shortening. The hydrolysis of ATP at a site on the cross-bridge supplies the chemical energy for shortening. The chemical energy from ATP hydrolysis must be transduced into mechanical energy at the actin binding site on the cross-bridge. The transduction of energy may involve the motion of domains within the cross-bridge. The specific modification of side chains on the cross-bridge with fluorescent or spin probes gives the opportunity to study the orientation of points on the cross-bridge and to test the proposed mechanisms for muscle contraction. The probe studies, although at first appearing to give contradictory impressions concerning cross-bridge rotation, now are shown to give a convergent picture that is consistent with the cross-bridge rotating during contraction. The basic model of energy transduction, involving a rigid body rotation of the cross-bridge, accounts for the orientation changes of the fluorescent and spin probes. The more refined mechanism, involving segmental mobility within the cross-bridge, is a possibility requiring further investigation.

Acknowledgment
This work was supported by grants from the National Science Foundation, U.S.A. (DMB-8819755), the National Institutes of Health, U.S.A. (1 R01 AR 39288-01A2), and the Mayo Foundation. T.P.B. is an Established Investigator of the American Heart Association.

REFERENCES

1. Huxley, A.F. "Reflections on Muscle," (1980). Princeton University Press, Princeton.
2. Huxley, H.E. *Science*, **164**, 1356 (1969).
3. Huxley, A.F. and Simmons, R.M. *Nature*, **233**, 533 (1971).
4. Abragam, A. and Bleaney, B. "Electron Paramagnetic Resonance of Transition Ions," (1970). Dover, New York.
5. Berliner, L.J. (ed.) "Spin Labeling: Theory and Applications," Vol. 1, (1976). Academic Press, New York.
6. Burghardt, T.P. and French, A.R. *Biophys. J.*, **56**, 525 (1989).
7. Burghardt, T.P. *Biopolymers*, **23**, 2383 (1984).
8. Ajtai, K. and Burghardt, T.P. *Biochemistry*, **28**, 2204 (1989).
9. Graceffa, P. and Seidel, J.C. *Biochemistry*, **19**, 33 (1980).
10. Laemmli, U.K. *Nature*, **227**, 680 (1970).
11. Ajtai, K., Pótó, L., and Burghardt, T.P. *Biochemistry*, **29**, 7733 (1990).
12. Reisler, E. *In* "Methods in Enzymology," Vol. 85, ed. D.W. Frederiksen and L.W. Cunningham, p. 84 (1982). Academic Press, New York.
13. Nihei, T., Mendelson, R.A., and Botts, J. *Biophys. J.*, **14**, 236 (1974).
14. Crowder, M.S. and Cooke, R. *J. Muscle Res. Cell Motil.*, **5**, 131 (1984).
15. Borejdo, J. and Putnam, S. *Biochim. Biophys. Acta*, **459**, 578 (1977).
16. Reisler, E., Burke, M., and Harrington, W. *Biochemistry*, **13**, 2014 (1974).
17. Miyanishi, T. and Borejdo, J. *Biochemistry*, **28**, 1287 (1989).
18. Burghardt, T.P. and Ajtai, K. *In* "Molecular Mechanisms in Muscular Contraction," ed. J.M. Squire, p. 211 (1990). Macmillan Press, London.
19. Quinlivan, J., McConnell, H.M., Stowring, L., Cooke, R., and Morales, M.F. *Biochemistry*, **8**, 3644 (1969).
20. Thomas, D.D. and Cooke, R. *Biophys. J.*, **32**, 891 (1980).
21. Cooke, R., Crowder, M.S., and Thomas, D.D. *Nature*, **300**, 776 (1982).
22. Ajtai, K., French, A.R., and Burghardt, T.P. *Biophys. J.*, **56**, 535 (1989).
23. Seidel, J.C. *In* "Methods in Enzymology," Vol. 85, ed. D.W. Frederiksen and L.W. Cunningham, p. 594 (1982). Academic Press, New York.
24. Aronson, J.F. and Morales, M.F. *Biochemistry*, **11**, 4517 (1969).
25. Nihei, T., Mendelson, R.A., and Botts, J. *Proc. Natl. Acad. Sci. U.S.A.*, **71**, 274 (1974).
26. Borejdo, J., Assulin, O., Ando, T., and Putnam, S. *J. Mol. Biol.*, **158**, 391 (1982).
27. Burghardt, T.P., Ando, T., and Borejdo, J. *Proc. Natl. Acad. Sci. U.S.A.*, **80**, 7515 (1983).
28. Ajtai, K. and Burghardt, T.P. *Biochemistry*, **25**, 6203 (1986).
29. Ajtai, K. and Burghardt, T.P. *Biochemistry*, **26**, 4517 (1987).
30. Botts, J., Thomason, J.F., and Morales, M.F. *Proc. Natl. Acad. Sci. U.S.A.*, **86**, 2204 (1989).

Mechanism of Actin Mediated Stimulation of the Myosin ATPase Reaction

AKIO INOUE,[*1] TOSHIAKI ARATA,[*1]
MITSUKUNI YASUI,[*2] MASATO OHE,[*3] AND
SHIN MURAI[*1]

*Department of Biology, Faculty of Science, Osaka University, Toyonaka, Osaka 560, Japan,[*1] Department of Chemical Engineering, Muroran Institute of Technology, Muroran, Hokkaido 050, Japan,[*2] and Department of Biochemistry, Dokkyo University School of Medicine, Mibu, Tochigi 321-02, Japan[*3]*

The function of heart as a pump depends on muscle which contracts by ATP hydrolysis. Cardiac muscle is composed of two kinds of filament, myosin and actin, as is true of skeletal muscle, and muscle contraction occurs as a result of the sliding of these filaments past each other. ATP is hydrolyzed by the head portion of the myosin molecule, and the heads of myosin form the crossbridges between myosin and actin filaments. Tension is developed by these crossbridges. The very low activity of myosin ATPase is dramatically accelerated by F-actin, and muscle contraction occurs only under conditions in which a high activity of actomyosin-type ATPase is expressed. The efficiency of energy transduction from chemical energy of ATP hydrolysis to mechanical work is extremely high.

The mechanism of coupling of ATP hydrolysis with muscle contraction has been studied based on the elementary steps of actomyosin ATPase reaction and also analysis of the reaction in which binding and dissociation occur between actin and myosin heads. We will first discuss the structure of myosin, especially the function of the two heads of myosin, the mechanism of actomyosin ATPase reaction, and finally the molecular mechanism of muscle contraction based on the mechanism of actomyosin ATPase reaction. Details of the structure of myosin and actin and the mechanism of actomyosin ATPase reaction have been described in review articles (*1-5*).

I. STRUCTURE OF MYOSIN AND THE MECHANISM OF MYOSIN ATPase
REACTION

Myosin plays the most important role in muscle contraction, and has
three functions: (1) formation of thick filaments, (2) interaction with
F-actin, and (3) hydrolysis of ATP. Figure 1 shows the structure of the
myosin molecule. Myosin has two separate heads with a long tail; both
heads can bind with actin and interact with ATP. The tail portion of the
myosin molecule (light meromyosin, LMM) forms the thick filament. The
two heads and LMM are connected by subfragment 2 (S-2). Therefore,
S-2 acts like the crankshaft of an engine.

Why does myosin have two heads and what is the function of each
head in muscle contraction? This is the most important question for
elucidation of the molecular mechanism of muscle contraction. We
showed that the ATPase reactions of these two heads differ (*1, 2*). One
(head A) forms the myosin-ATP complex and the other (head B) forms
the myosin-P-ADP complex (M-P-ADP) as a stable intermediate. The
value of K_m for the ATPase reaction of head B is much lower than that
of head A. Since the rate of decomposition of the myosin-P-ADP complex
is greatly accelerated by F-actin, head B may catalyze the actomyosin-
type ATPase reaction which is coupled with muscle contraction. On the
other hand, head A is considered to work to increase the efficiency of
contraction.

Heads B and A have been separated using their different affinities
for ATP (*6, 7*). Both heads can bind with actin and interact with ATP,
but the actomyosin-type ATPase and EDTA-ATPase were catalyzed only

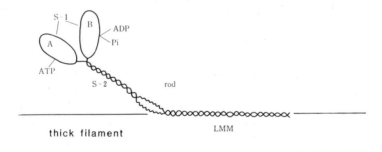

Fig. 1.   Schematic illustration of the structure of myosin molecule. Myosin has
two different heads, one (A) forms the myosin-ATP complex and the other (B)
forms the myosin-P-ADP complex.

by head B (6, 7). It was also shown that the two heads have different chemical structures (8, 9). The reactivity of a specific lysine residue is different between them, and more recently the amino acid sequence around a specific lysine residue (86 Lys) has also been found to be different. Recently, the two heads were separated using antibodies against peptides around this residue (S. Murai, T. Arata, T. Miyanishi, and A. Inoue, in preparation). It is thus expected that their function during contraction will soon be clarified.

The mechanism of myosin ATPase reaction has been studied by many researchers, and almost the same scheme has been proposed (1-3), though there is a diversity of opinions whether the two heads of myosin are identical or different from each other. In head B, M-P-ADP is formed *via* loosely-bound and tightly-bound myosin-ATP complexes (M_a-ATP and M_bATP, respectively). It was also shown that M_bATP is in rapid equilibrium with M-P-ADP. M-P-ADP decomposed slowly into M + ADP + P_1. Several reports proposed that M-P-ADP is decomposed *via* the M-ADP complex. However, M-ADP is not stable, and is formed transiently even if M-ADP exists, whereas in head A ATP is hydrolyzed *via* the myosin-ATP complex.

II.   MECHANISM OF ACTOMYOSIN ATPase REACTION

Muscle contraction is directly coupled with actomyosin-type ATPase reaction, so that to elucidate the molecular mechanism involved, it is important to understand the mechanism of actomyosin ATPase reaction. Since muscle fiber has highly ordered structure, and since the rate of ATP hydrolysis by F-actin-myosin complex depends greatly on the progression of the superprecipitation of actomyosin, the mechanism of actomyosin-type ATPase reaction was studied using acto-subfragment 1 (acto-S-1) or acto-heavy meromyosin (acto-HMM) in solution. It has generally been accepted that actomyosin-type ATPase reaction is catalyzed through the myosin-P-ADP complex which is the most stable ATPase intermediate in the myosin ATPase reaction.

When ATP is added to acto-HMM or acto-S-1 at high ionic strength, HMM or S-1 dissociates from F-actin. Dissociation of acto-HMM or acto-S-1 occurs by formation of myosin-ATP complex. The dissociation of actomyosin does not require the hydrolysis of ATP. M-ATP is rapidly transformed into M-P-ADP. The myosin head recombines with F-actin after the decomposition of M-P-ADP into M + ADP +

P_1. At low ionic strength, M-P-ADP can recombine with F-actin, and myosin returns to the original state. The simplest mechanism of acto-myosin ATPase reaction based on the above findings would be that ATP is hydrolyzed *via* AM-ATP, A+M-ATP, A+M-P-ADP, AM-P-ADP, and AM+ADP+Pi; however, the following results cannot be explained by this mechanism. We found that: (1) at low ionic strength and high concentration of F-actin S-1 does not dissociate from F-actin even in the presence of ATP, and (2) the rate of recombination of S-1-P-ADP complex with F-actin is lower than that of acto-S-1 ATPase reaction in the steady state. We therefore proposed that ATP is hydrolyzed *via* two routes: one with dissociation and recombination of acto-S-1 into F-actin + S-1-P-ADP (outer route), and the other without the dissociation of acto-S-1 (inner route) (*10, 11*) (see Fig. 2).

The rate of ATP hydrolysis *via* the outer route is given by (amount of dissociated S-1)×(the rate of recombination of M-P-ADP with F-actin), while the rate of the inner route is given by (amount of bound S-1)×(the rate of direct decomposition through AM-P-ADP). We have found that the rate of overall reaction (v) is given by $v=(1-\alpha) V+\alpha k_r$, where α, V, and k_r are the extent of dissociation of the myosin head from F-actin, the rate of ATPase reaction at infinite concentration of F-actin, and the rate constant of binding of M-P-ADP with F-actin. Thus, the above mechanism was supported by the kinetic analysis (*11*).

In muscle fibers the concentration of protein is extremely high, and

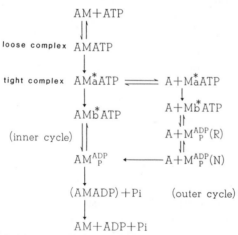

Fig. 2.   Mechanism of actomyosin ATPase reaction. R and N denote the re-fractory and non-refractory state, respectively.

the diffusion of ligand is insufficient for analysis of the rate of elementary steps of actomyosin ATPase reaction; this makes difficult to analyze the mechanism of ATP hydrolysis by muscle fibers during contraction. We analyzed the mechanism of actomyosin-type ATPase reaction in muscle fibers by an oxygen exchange reaction (*12, 13*). The ATPase reaction was carried out using ^{18}O-ATP as a substrate and the distribution of ^{18}O-species of Pi was measured using a gas-chromatograph mass-spectrometer. When ^{18}O-ATP is hydrolyzed into ADP + Pi, ^{16}O is incorporated into Pi at the step from M-ATP to M-P-ADP. However, if M-P-ADPi is transformed into M-ATP, another ^{16}O can be incorporated (Fig. 3). The distribution of Pi with 3, 2, 1, and no ^{18}O was determined by the distribution of ^{18}O in γ-position of ATP, the ratio of rate constant from M-P-ADP to M-ATP and that from M-P-ADP to M + ADP + Pi. Since the rate of decomposition of AM-P-ADP into AM + ADP + Pi in the inner cycle is extremely high, the extent of oxygen exchange should be low when ATP is hydrolyzed *via* the inner cycle. On the other hand, M-P-ADP in the outer cycle is in rapid equilibrium with M-ATP; therefore, the oxygen exchange during ATP hydrolysis *via* the outer cycle is considered to be very high. The distribution of ^{18}O-Pi species obtained after acto-S-1 ATPase reaction revealed that Pi was produced by two kinds of ATPase reaction: one with a low and the other with a high extent of ^{18}O exchange reaction. The ratio of ATPase reactions with low and high oxygen exchange was equal to the ratio of ATPase reactions through the inner and outer routes of this reaction measured by the kinetic method.

We examined the distribution of ^{18}O-Pi when ^{18}O-ATP was hydrolyzed by glycerol-treated muscle fibers. The distribution of ^{18}O-Pi species

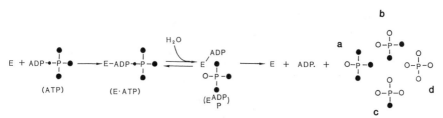

Fig. 3. Oxygen exchange reaction during the ATPase reaction. ^{16}O is incorporated into Pi at the step from M-ATP to M-P-ADP. When M-P-ADP is transformed into M-ATP, another ^{16}O is incorporated. Since the extent of oxygen exchange during the outer route of the ATPase reaction is high but that during the inner route of the ATPase reaction is low, we can determine the route of ATP hydrolysis from the distribution of ^{18}O-Pi species.

fitted well if the fraction of ATP hydrolyzed through the inner route of the ATPase reaction was 73–80% that of total ATP hydrolysis (*13*). This result suggested that muscle contraction is coupled with the inner route of the reaction.

The elementary step of actomyosin ATPase reaction *via* the inner cycle has been studied using acto-S-1 at very low ionic strength or crosslinked acto-S-1 (*14*). It has shown that ATP is hydrolyzed through loose acto-S-1-ATP complex, tight acto-S-1-ATP complex, and acto-S-1-P-ADP complex. The values of K_m and V_{max} for overall reaction were 12 μM and 13 sec^{-1}, respectively. The rate-determining step of ATPase was the step from tight acto-S-1-ATP complex to acto-S-1-P-ADP complex rather than the decomposition of acto-S-1-P-ADP complex. We calculated the free energy change in the elementary steps of acto-S-1-ATPase reaction from the equilibrium constant of each step, and found that most of the energy obtained by ATP hydrolysis was liberated at the step from acto-S-1-P-ADP to acto-S-1-ADP complex. Therefore, this step may be coupled with the development of tension by muscle fibers.

III.   MECHANISM OF MUSCLE CONTRACTION

It was suggested that muscle contraction occurs as a result of the following three reactions: (1) formation of crossbridges, (2) movement of crossbridges which results in the development of tension, and (3) decomposition of crossbridges. In muscle fibers ATP is hydrolyzed *via* the inner route of the ATPase reaction. In this route, only acto-S-1-ATP is in rapid equilibrium with the dissociated form of acto-S-1 (F-actin + S-1-P-ADP). Therefore, steps (1) and (3) may occur as a result of this step. Since most of the energy obtained by ATP hydrolysis is liberated at the step from AM-P-ADP to AM-ADP or AM, we considered that the development of tension is coupled with the step from AM-P-ADP to AM-P-ADP. This idea agrees with the result of Arata (*15*) that the distance between specific thiol residues in S-1 and actin differs between acto-S-1-ATP, acto-S-1-ADP and acto-S-1-ADP, acto-S-1.

Figure 4 shows the molecular mechanism of muscle contraction based on these results. In this model, it is assumed that movement of the head occurs by changing the tilting angle of myosin head bound to F-actin, and the elastic component is located in the junction between S-2 and LMM. Energy liberated by ATP hydrolysis is stored in the elastic component, and sliding of the filament occurs when the elastic compo-

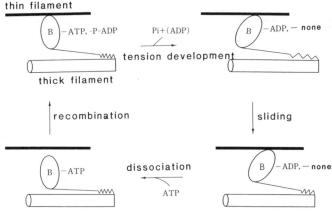

Fig. 4.   Mechanism of muscle contraction.

nent is shortened. After liberation of the energy stored in the elastic component, the dissociation of crossbridges occurs by formation of acto-S-1-ATP complex, and the myosin head then recombines with another actin molecule.

The Two Heads of Myosin

As stated in section I, myosin has two separate heads which are connected by a tail, and it is considered that the movement of one head may disturb the other head of the molecule. It is difficult to believe that these two heads are identical and yet interact independently with F-actin, but Fig. 5 shows how this interaction occurs. F-actin is known to be a two stranded helical polymer of G-actin and both strands have the same polarity. Since the two myosin heads have similar structures, it is difficult to understand how they could bind from either side of F-actin. From the arrangement of myosin and actin filaments in muscle fibers, it is also difficult to understand how two heads interact with different actin filaments. The most feasible mechanism is that they interact with F-actin from the same side of the actin filament. In this model the two heads must interact like stepping worm.

It has been shown that the two heads of myosin are different and that only one of them catalyzes the actomyosin ATPase reaction. If the interaction between head A and F-actin is regulated by the steps of ATPase reaction in head B, we can draw a highly efficient model of muscle contraction. Dissociation of myosin from F-actin is induced by one

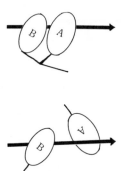

Fig. 5.   Binding of the two heads of myosin with actin filament.

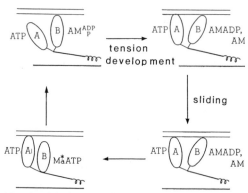

Fig. 6.   Mechanism of muscle contraction by the two heads of myosin.

mole of ATP per mole of myosin, *i.e.*, by formation of M-P-ADP complex. Thus, the affinity of head A with F-actin may be low when the other head (B) forms M-P-ADP. We assumed that head A dissociates from F-actin when head B forms AM-P-ADP, and that head A binds with F-actin when head B is at the state of AM and A-M-ATP. Figure 6 shows a model of contraction based on these assumptions. When head B moves on F-actin by transition from A-M-P-ADP to A-M-ADP, head A dissociates from F-actin and does not disturb the movement of head B. When head B dissociates from F-actin and then recombines with F-actin, head A binds with F-actin and prevents the slip of crossbridges.

SUMMARY

Muscle contraction occurs by a cyclic reaction of myosin, actin and

ATP. We have studied the mechanism of ATP hydrolysis by acto-S-1 and the binding of myosin heads with F-actin during the elementary steps of this hydrolysis. Oxygen exchange analysis was used in studies of the mechanism of ATP hydrolysis by muscle fibers, and we found that ATP is hydrolyzed *via* two routes, one with and the other without the accompanying dissociation of acto-S-1 into F-actin+S-1-P-ADP. In muscle fibers, ATP is mainly hydrolyzed *via* the direct decomposition of acto-S-1-P-ADP complex. The movement of crossbridges and development of tension was suggested to be coupled with the decomposition of acto-S-1-P-ADP into acto-S-1-ADP. Dissociation and reformation of crossbridges (acto-S-1) may occur by formation of the acto-S-1-ATP complex. We also showed that the two heads of myosin have different structures and functions, one of the two forming the S-1-P-ADP complex. We proposed the mechanism of muscle contraction induced by the two heads of the myosin molecule.

REFERENCES

1. Inoue, A., Takenaka, H., Arata, T., and Tonomura, Y. *Adv. Biophys.*, **13**, 1 (1979).
2. Inoue, A. and Tonomura, Y. *Mol. Cell. Biochem.*, **5**, 127 (1974).
3. Taylor, E.W. *CRC Crit. Rev. Biochem.*, **6**, 103 (1979).
4. Sugi, H. and Pollack, G.H. (eds.) *"Molecular Mechanism of Muscle Contraction,"* (1989). Plenum Press, New York.
5. Paul, R.J., Elzinga, G., and Yamada, K. (eds.) *"Muscle Energetics,"* (1989). Alan R. Liss Inc., New York.
6. Inoue, A. and Tonomura, Y. *J. Biochem.*, **79**, 419 (1976).
7. Inoue, A. and Tonomura, Y. *J. Biochem.* **80**, 1359 (1976).
8. Miyanishi, T., Inoue, A., and Tonomura, Y. *J. Biochem.*, **85**, 747 (1979).
9. Miyanishi, T., Maita, T., Matsuda, G., and Tonomura, Y. *J. Biochem.*, **89**, 1845 (1979).
10. Inoue, A., Ikebe, M., and Tonomura, Y. *J. Biochem.* **88**, 1663 (1980).
11. Inoue, A., Shigekawa, M., and Tonomura, Y. *J. Biochem.*, **74**, 923 (1973).
12. Yasui, M., Ohe, M., Kajita, A., Arata, T., and Inoue, A. *J. Biochem.*, **104**, 644 (1988).
13. Yasui, M., Ohe, M., Kajita, A., Arata, T., and Inoue, A. *J. Biochem.*, **105**, 665 (1989).
14. Inoue, A., Arata, T., and Yasui, M. *In "Muscle Energetics,"* ed. R.J. Paul, G. Elzinga, and K. Yamada, p. 15 (1989). Alan R. Liss Inc., New York.
15. Arata, T. *J. Mol. Biol.*, **164**, 107 (1986).

Molecular Mechanisms for Cardiac Cellular Hypertrophy Due to Mechanical Stimuli

YOSHIO YAZAKI, ISSEI KOMURO, EITETSU HOH, AND
RYOZO NAGAI

*Third Department of Internal Medicine, Faculty of Medicine, University of Tokyo, Tokyo
113, Japan*

During the process of cardiac hypertrophy, the expression of specific genes such as protooncogenes and fetal-type genes of contractile proteins was induced and there was also an increase in protein synthesis. The "immediate early genes" like protooncogenes and heat shock protein genes are induced as an early response to pressure overload, and "late responsive genes", fetal contractile protein genes and the atrial natriuretic peptide gene are reexpressed as a later event. Many hormones and growth factors have recently been reported to induce cardiac hypertrophy and specific gene expression in cultured cardiac myocytes. However, whether hemodynamic overload directly stimulates cellular hypertrophy and specific gene expression in cardiac myocytes without the participation of humoral factors remains unknown.

To examine whether mechanical stimuli are directly coupled to specific gene expression, we cultured neonatal rat cardiac myocytes in deformable silicone culture dishes with defined serum-free medium, imposed mechanical stimuli by stretching adherent myocytes, and examined protooncogene expression. Further, using protooncogene induction by stretching, we studied the signal transduction pathway of mechanical stimuli on cardiac myocytes. The activation of MAP kinase which is proposed to activate S6 kinase of ribosomes resulting in increased efficiency of protein synthesis was also of interest.

I. MECHANICAL LOAD STIMULATES CELL HYPERTROPHY AND SPECIFIC
GENE EXPRESSION

Deformable culture dishes were devised entirely of silicone to impose
mechanical stimuli directly on cardiac myocytes. The bottom of the dish
was 1-mm thick, and the whole was highly transparent because no
inorganic filler was used in either component. Dishes were mechanically
expanded using a plastic frame and their length increased uniaxially,
then the attached cardiac myocytes were stretched. The resting length of
the myocytes was increased parallel to the axis of expansion by the same
percent length as the dish (1). This method allowed a more detailed
analysis including quantitative assessment of mRNA levels because
samples of larger scale were obtained.

Primary cultures of cardiac myocytes were prepared from the
ventricles of 1-day-old Wistar rats. A cardiac myocyte-rich fraction was
obtained by preplating method. Myocytes not attached to the replated
dishes were plated into laminine-coated silicone dishes at a field density
of 1×10^5 cells/cm^2. A nonmuscle cell-rich fraction was obtained by
preplating the cells into silicone dishes for the first hour.

The effect of myocyte stretch on amino acid incorporation into
cardiac proteins is shown in Fig. 1A. To avoid the effect of serum, this
experiment was performed after 2 days in the serum-free, chemically
defined medium. Myocytes were stimulated by 10% increase in the length
of the attached dishes. At this point, more than 90% of the cells were
beating. The incorporation of [^{14}C]phenylalanine was significantly increased 2 hr after stretch and the stimulation was maintained for over 12
hr (Fig. 1A), suggesting that mechanical stress stimulates cardiac cellular
hypertrophy.

To ascertain whether mechanical stress induces specific genes such
as protooncogenes and fetal-type isogenes of contractile proteins as
observed in the heart *in vivo*, we examined the expression of c-*fos* and
skeletal α-actin gene. Northern blot analysis revealed that c-*fos* was
rapidly and transiently expressed by stretching myocytes. The level of
c-*fos* mRNA was increased as early as 15 min after the passive stretch of
myocytes, and reached the maximum level at 30 min, followed by a
decline to an undetectable level (Fig. 1B). The kinetics of this c-*fos*
expression by stretching is the same as those when cells are stimulated by
serum or growth factors. This protooncogene expression was more

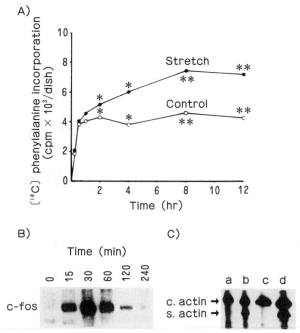

Fig. 1.   Stimulation of amino acid uptake and c-*fos* and skeletal α-actin gene expression by myocyte stretching. A: after 2 days in a serum-free medium, culture dishes were stretched by 10% in length along single axis and 1 μCi/ml [^{14}C]-phenylalanine was added for 30 min prior to processing the cells for intracellular trichloroacetic acid-insoluble radioactivity. Each point represents the mean±S.E. of three experiments performed in duplicate. *$p<0.05$; **$p<0.01$. B: cardiac myocytes were stretched by 10% for 30 min. RNA was extracted and 10 or 20 μg of total RNA (indicated in parentheses) was analyzed by Northern blot hybridization using a 0.8-kb AccI fragment of human c-*fos* as a probe. C: cardiac α-actin (c. actin) and skeletal α-actin (s. actin) were separated by a primer extension technique. RNA was extracted from neonatal cardiocytes cultured for 1 (a, b) or 2 (c, d) days with (b, d) or without (a, c) stretching.

abundant in cardiac myocyte-rich fraction than in nonmuscle cell-rich fraction, confirming that the stimulation of c-*fos* gene expression by stretching occurred in cardiac myocytes. The induction of c-*fos* mRNA depends on the extent of expansion of the dishes. The stimulation of c-*fos* gene expression was recognized by a 5% increase in the length of the dishes, and maximum stimulation was obtained by 20% of stretch.

The level of skeletal α-actin mRNA was also elevated after the passive stretch of myocytes. Skeletal α-actin mRNA was significantly increased 4 hr after stretching, and gradually accumulated for up to 2

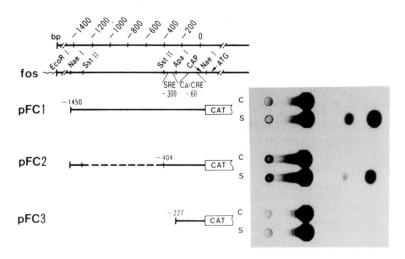

Fig. 2. CAT activity of c-*fos* recombinants containing deletions in the 5′-flanking region. Cardiac myocytes were transfected with 1 μg of DNA per dish and stretched for 2 hr. Cell extracts were prepared and assayed for CAT activity. C, control cells; S, stretched cells; SRE, serum response element; Ca/CRE, calcium/ cAMP response element.

days during stimulation (Fig. 1C). Because acute pressure overload is known to induce cardiac hypertrophy and gene expression such as protooncogenes and fetal-type isogenes of contractile proteins in the heart *in vivo*, our findings on the expression of c-*fos* and skeletal α-actin gene by myocyte stretching suggested that stretching cardiac myocytes *in vitro* could substitute for hemodynamic overload *in vivo*.

II. TRANSCRIPTIONAL ACTIVATION OF c-FOS GENE BY MYOCYTE STRETCHING

To determine whether the c-*fos* gene expression by myocyte stretching was regulated at the transcriptional level or the post-transcriptional level, we analyzed its promoter function by chloramphenicol acetyltransferase (CAT) assay method (2). Linking the 5′ flanking region of the *fos* gene including its promoter to the 5′ end of the CAT encoding sequences in the plasmid, the plasmid was transfected into primary cultures of neonatal rat cardiac myocytes and the CAT activity of the cell extracts measured. The pSVO CAT construct containing the entire coding sequences of the procaryotic CAT gene minus its promoter showed very little CAT activity in either the absence or presence of stretch stimulation. By contrast, when pSVO *fos* CAT, which contained the 5′ c-*fos* flanking region, was introduced into the system, myocyte stretching for 48 hr reproducibly caused more than a 7-fold increase in CAT activity, although there was little activity without stretching. When the pSV2CAT construct, which contained SV40 enhancer and early promotor sequences, was introduced into myocytes, a great amount of CAT activity was observed; no additional activity, however, was obtained after stretching. Furthermore, a run-on study using myocyte nuclei also revealed the accumulation of c-*fos* mRNA by stretching. These results suggested that the c-*fos* gene expression caused by stretching was regulated at the transcriptional level and that the stretch response element was located in the 5′ flanking region of the gene.

Mechanical stress thus markedly induced the expression of c-*fos* protooncogenes without the participation of humoral factors. Therefore, hemodynamic overload itself seems to be one of the main factors stimulating the expression of the c-*fos* gene in the heart *in vivo*. Fos, the protein which the *fos* gene encodes, was recently found to be being localized in the nucleus and to bind to the 12-*O*-tetradecanoylphorbol-13-acetate (TPA)-responsive elements of some genes, followed by the activation of their gene transcription in cooperation with the transcription factor AP-1. These observations suggest that some early responsive gene products like Fos may stimulate their subsequent gene expressions in the heart under conditions of hemodynamic overload.

III. MECHANISM OF c-*FOS*-GENE EXPRESSION BY MYOCYTE STRETCHING

To identify the sequences essential for the transcription of the *fos* gene induced by stretching, we analyzed effects of deletion of the 5′ flanking region of the gene on CAT activity.

Deletion mutagenesis of the 5′ flanking region of the c-*fos* gene indicated that the sequences between -227 and -404 base pairs were required for the efficient transcription of the *fos* gene by myocyte stretching (Fig. 2) (*2*). Since serum and cAMP responsive element are known to be in this region, we hypothesized that three factors, cAMP, protein kinase C, and tyrosine kinase are involved in c-*fos* gene stimulation by stretching.

A desensitization study was carried out to determine which protein kinase system plays the central role in the signal transduction induced by

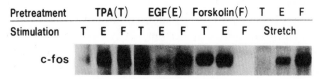

Fig. 3. Desensitization, down-regulation, and pharmacological study of c-*fos* induction. A: myocytes were pretreated with 10 ng/ml TPA (T), 10 ng/ml EGF (E), or 1 μM forskolin (F) for 4 hr, then stimulated by the same doses of these inducers or by stretching for 30 min. B: cells were stretched in the presence and absence of the indicated inhibitors for 30 min. RNA was extracted, run on a gel, and transferred to the filters. The filters were hybridized with the c-*fos* probe. S, stretched cells.

mechanical stress. Myocytes were pretreated with either phorbol esters (TPA), epidermal growth factor (EGF) or forskolin for 24 hr to down-regulate individual protein kinases, C kinase, tyrosine kinase or A kinase, respectively, and then treated again with one of these inducers. Thirty min after the treatment, we assessed the mRNA level of c-*fos* by Northern blot analysis. Following TPA pretreatment, no c-*fos* stimulation was found. Pretreatment with EGF or forskolin induced desensitization against the stimulation with EGF or forskolin, respectively. The most important observation was that only the pretreatment with TPA desensitized myocytes against stretch (Fig. 3). The induction of c-*fos* gene by myocyte stretching might, therefore, be caused by the activation of protein kinase C.

To confirm this possibility, the effect of protein kinase C inhibitors on the expression of c-*fos* by myocyte stretching was examined. H-7 strongly inhibited c-*fos* mRNA induction by stretching, whereas H-1004 inhibited it weakly, depending on the K_i value for protein kinase C. Staurosporin also strongly inhibited c-*fos* induction. The treatment of myocytes with TPA induced both c-*fos* and skeletal α actin mRNA.

Finally, immediately after stretching, the activation of phosphatidyl inositol turnover was observed in myocytes to determine the mechanism activating protein kinase C. One min after stretching, inositol monophosphate and bisphosphate significantly increased and after 5 min reached about 2-fold the level of control (Fig. 4). No elevation in inositol trisphosphate levels was recognized at either time point. However, these results suggest that mechanical stress might stimulate protein kinase C activity *via* phospholipase C activation in cardiac myocytes. Recently, mechanical

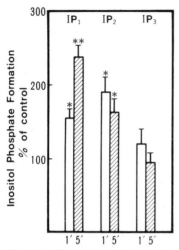

Fig. 4.   Effect of stretching on the level of inositol phosphates. Myocytes were incubated for 24 hr in Dulbecco's modified Eagle's medium containing myo-[2-^{3}H]inositol and stretched 10% for the indicated times. The water-soluble inositol phosphates were separated by column chromatography. Each histogram represents the average percentage of control from five experiments performed in duplicate (mean $\pm$ S.E.). Control values (cpm $\times$ 10^{-2}/dish) were as follows: inositol monophosphate (IP$_1$), 160–350; inositol bisphosphate (IP$_2$), 30–42; inositol tris-phosphate (IP$_3$), 62–93. Statistical significance was determined by analysis of variance (*39*). *$p < 0.05$; **$p < 0.01$.

stress has been reported to induce prostaglandin production in skeletal and endothelial cells *via* phospholipase C pathway; these reports might support our hypothesis. However, further investigation is necessary to elucidate the precise molecular mechanism(s) by which mechanical load activates phospholipase C or protein kinase C (*2*).

IV.   ACTIVATION OF MAP KINASE BY MYOCYTE STRETCHING

Mitogen-activated protein (MAP) kinase is a serine/threonine pro-tein kinase which can be activated by a variety of stimuli such as growth factors and TPA, and is proposed to be a general intracellular signaling molecule linking events at the cell surface, in the nucleus and in the ribosome. Recent reports indicated that MAP kinase can phosphorylate and activate S6 kinase of ribosomes resulting in increased efficiency of protein synthesis (*3, 4*). Thus, in order to further define the intracellular signals by which mechanical stimuli lead to increased protein synthesis, we examined whether mechanical stimuli can activate MAP kinase.

Cardiac myocytes were labeled with ^{32}P orthophosphate and stretched, solubilized and immunoprecipitated with specific antibodies against MAP kinase. Stretching myocytes increased phosphorylation of the protein of 43 kDa by approximately 6-fold, while other bands were not increased. This protein was identified as the 43 kDa MAP kinase, MAP II kinase by Western immunoblotting.

Kinase activity of cell lysate was measured by the phosphorylation of myelin basic protein (MBP) as substrate to learn whether increased phosphorylation of MAP II kinase by stretching is indeed associated with increased kinase activity. The phosphorylation of MBP was stimulated maximally approximately 1.8-fold 10–20 min after stretching (Fig. 5).

Determination of whether the phosphorylation of MBP is induced by MAP kinase itself was made by electrophoresing cell lysates on SDS-polyacrylamide gels containing MBP. After renaturation, the gels were incubated with γ-^{32}P ATP and magnesium. We observed an increased phosphorylation of MBP at 43 kDa after stretching; this increase was also noted when myocytes were treated with TPA. These data demonstrate that stretching myocytes increased 43 kDa MAP kinase

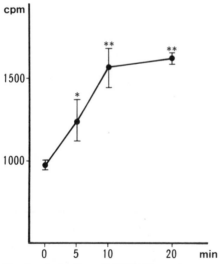

Fig. 5. Activation of MAP kinase by myocyte stretching. Activities of MAP kinase were measured in cell lysates with the phosphorylation of MBP as substrate. Maximal stimulation of MBP phosphorylation (approximately 1.8-fold) occurred 10–20 min after stretching. *$p < 0.01$; **$p < 0.005$.

(MAP II kinase) activity. The phosphorylation of MBP was also recognized at 70 kDa, but the identity of this kinase activity is unknown at present.

Our results strongly suggested that MAP kinase is stimulated by mechanical stress *via* protein kinase C activation, and this activated MAP kinase probably phosphorylates and stimulates S6 kinase of ribosomes accompanied by an increased protein synthesis. MAP kinase has been shown to phosphorylate and activate the protooncogene product, c-Jun. Therefore, the induction of c-*fos* and possible instigation of c-*jun via* MAP kinase activation by stretching can synergistically activate the function of AP1 complex.

SUMMARY

To examine the molecular mechanisms by which mechanical stimuli induce protooncogene expression, we cultured rat neonatal cardiocytes in deformable dishes and imposed an *in vitro* mechanical load by stretching the adherent cells. Myocyte stretching increased total cell RNA content and mRNA levels of c-*fos* and skeletal α-actin followed by activation of protein synthesis. CAT assay indicated that sequences containing a serum response element were required for efficient transcription of c-*fos* gene by stretching. This accumulation of c-*fos* mRNA was suppressed by protein kinase C inhibitors at the transcriptional level and inhibited markedly by down-regulation of protein kinase C. Moreover, myocyte stretching increased inositol phosphate levels. These findings suggest that mechanical stimuli might directly induce protooncogene expression, possibly *via* protein kinase C activation. We also observed the activation of MAP kinase by myocyte stretching. This may indicate that MAP kinase activation increases in efficiency of protein synthesis on ribosomes induced by mechanical stimuli.

Acknowledgment

This investigation was supported in part by a Grant-in-Aid for Scientific Research from the Ministry of Education, Science and Culture of Japan, Grants were also received from Japan's Ministry of Welfare and the Uehara Memorial Foundation, and for Basic Research on Cardiac Hypertrophy from the Vehicle Racing Commemorative Foundation.

REFERENCES

1. Komuro, I., Kaida, T., Shibazaki, Y., Kurabayashi, M., Katoh, Y., Hoh, E., Takaku F., and Yazaki, Y. *J. Biol. Chem.*, **265**, 3595 (1990).
2. Komuro, I., Katoh, Y., Kaida, T., Shibazaki, Y., Kurabayashi, M., Hoh, E., Takaku, F., and Yazaki, Y. *J. Biol. Chem.*, **266**, 1265 (1991).
3. Ahn, N.G. and Krebs, E.G. *J. Biol. Chem.*, **265**, 11495 (1991).
4. Ahn, N.G., Seger, R., Bratlien, R.L., Diltz, C.D., Tonks, N.K., and Krebs, E.G. *J. Biol. Chem.*, **266**, 4220 (1991).

Mechanism of Mitochondrial Dysfunction during Ischemia and Reperfusion

TORU INOUE,[*1] MOTONOBU NISHIMURA,[*2]
AND KUNIO TAGAWA[*1]

*Department of Physiological Chemistry[*1] and First Department of Surgery,[*2] Medical School, Osaka University, Suita, Osaka 565, Japan*

When a tissue is deprived of its blood supply, the cellular levels of high-energy compounds decrease, resulting in many metabolic disturbances. On reperfusion after brief ischemia, cells regenerate ATP and recover physiological functions, but prolonged ischemia leads to irreversible injuries and cell death. The way in which reversible injuries become irreversible is not known exactly, but naturally mitochondrial dysfunction is one of the most serious events implicated in ischemic cell death because mitochondria are responsible for oxidative phosphorylation. We have found that mitochondria in rat liver lose their phosphorylating capacity after 120 min of ischemia (*1*), and that this period exactly coincides with that for loss of reversibility of the cellular ATP level in ischemic liver (*2*). Several mechanisms have been proposed for the mitochondrial dysfunction: inhibition of ATP/ADP translocase by long-chain acyl-CoA, which accumulates during anoxia (*3*); decrease in cytochromes (*4*) and phospholipids in the mitochondrial membrane (*5*), and overload of Ca^{2+} (*6*). An alteration of cellular Ca^{2+} homeostasis is proposed to play a central role in various types of cell injuries including ischemia (*7*). Since a 10,000-fold gradient of free Ca^{2+} concentration across the plasma membrane is maintained by energy-dependent Ca^{2+}-extruding systems (*8*), anoxia is thought to lead to cytoplasmic and mitochondrial Ca^{2+}-loading. An elevated Ca^{2+} level in the cytosol will activate several kinds of Ca^{2+}-dependent hydrolytic enzymes such as phospholipases and proteases, resulting in breakdown of cell structures. Moreover, excess

Ca^{2+}-loading of isolated mitochondria is reported to result in uncoupling of oxidative phosphorylation (9).

"Reperfusion injury" is another phenomenon implicated in the irreversibility of ischemic damage (10). Reperfusion is essential for recovery from ischemic injuries, but cells reversibly injured at the end of an ischemic period are often injured more severely, or irreversibly during reperfusion. The formation of oxygen free radicals upon reoxygenation of anoxic tissues has been thought to be the most important factor in this phenomenon. Several sources of oxygen radicals in ischemia-reperfused heart have been proposed: xanthine oxidase formed from xanthine dehydrogenase by a Ca^{2+}-dependent protease during ischemia (11); activated leukocytes, which migrate into ischemic myocardium during reperfusion (12), and the electron transport system in mitochondria (13). The former two are, however, involved in the late phase, after reperfusion for more than an hour, and are not associated with the reversibility in the early phase which we discuss here.

In this paper, we review our recent studies on mitochondrial dysfunction associated with ischemia and reperfusion. During ischemia, Ca^{2+} is released from anoxic mitochondria rather than being taken up. This released Ca^{2+} increases the cytoplasmic concentration of Ca^{2+}, resulting in loss of phosphorylation capacity. Upon reperfusion, reoxygenated mitochondria are also severely damaged by a completely different mechanism. Mitochondria accumulate Ca^{2+} concomitantly with respiration, and then their membrane becomes leaky to various substances including matrix proteins.

I. ANOXIC INJURY OF MITOCHONDRIA

The respiratory control ratio of mitochondria isolated from ischemic liver decreases in proportion to the duration of ischemia, and the decrease has been found to be due mainly to increased H^+ permeability of the inner membrane (1). As shown in Fig. 1, when intact mitochondria are incubated under anoxic conditions, their ATP content decreases, and their respiratory control ratio also decreases, as observed in ischemic liver. After 120 min of anoxia, the ratio becomes equal, indicating complete loss of the diffusion barrier against H^+. Figure 1 also shows that matrix Ca^{2+} is released into the external medium, and that non-esterified polyunsaturated fatty acids are liberated from membrane phospholipids. Since the amounts of mono-unsaturated and saturated fatty acids liberat-

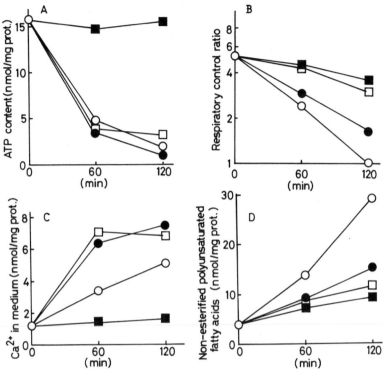

Fig. 1.   Effects of ATP analog and EGTA on anoxic dysfunction of mitochondria. Mitochondria isolated from rat liver were incubated under anoxic conditions with three kinds of agents. At the times indicated, the intramitochondrial concentrations of ATP and its analog (A), the respiratory control ratio (B), the concentration of Ca^{2+} in the medium (C), and the amount of non-esterified polyunsaturated fatty acids (D) were measured. For further details, see ref. *14*. ○ no addition; ● 0.5 mM quinacrine; □ 2 mM EGTA; ■ 3 mM β,γ-methylene adenosine 5'-triphosphate.

ed are small (*14*), activation of mitochondrial membrane-bound phospholipase A_2 is suggested. Addition of a non-metabolizable ATP analog which is rapidly incorporated into mitochondria (*15*), prevents all these changes, indicating that the depletion of ATP is the initial event in the sequence of a process induced by anoxia. A chelator of Ca^{2+}, EGTA and inhibitors of phospholipase A_2 prevent the liberation of fatty acids and impairment of the phosphorylating capacity without affecting the depletion of ATP or the release of Ca^{2+}. These results indicate that disruption of the barrier against H^+ is due to hydrolysis of membrane phospholipids by phospholipase A_2, which is activated by released Ca^{2+}. Although phospholipase A_2 activated in the presence of large amounts of Ca^{2+} may induce increase in non-specific permeability of the inner membrane (discussed below), anoxic mitochondria at this stage do not lose the diffusion barrier against substances other than H^+, such as adenine nucleotides and matrix enzymes (1, 16).

II. REOXYGENATION INJURY OF PERFUSED RAT HEART

Reintroduction of oxygen to anoxic heart is reported to induce massive leakage of myocardium enzymes (17) and striking morphological changes (18), both of which are much more marked than those found in the same period of anoxia only. This phenomenon, known as the oxygen paradox, has been considered to be caused by oxygen radicals generated in the early phase of reoxygenation. We have found, however, that resumption of cardiac beating on reoxygenation also plays a significant role in enzyme leakage (19). Figure 2 shows that the leakage of aspartate aminotransferase (AST) from perfused rat heart is markedly accelerated on resumption of beating rather than K^+-arrested reoxygenation. Arti-

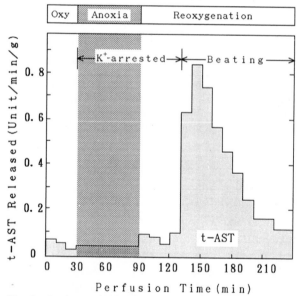

Fig. 2.   Leakage of AST from a perfused rat heart during anoxia and reoxygenation. During anoxia, cardiac beating was arrested by perfusion with high-K^+ medium, and spontaneous beating was resumed on switching to normal medium after 40 min of reoxygenation. For further details, see ref. 19.

ficial mechanical stress on the left ventricular wall during prolonged anoxia, like reoxygenation, provokes leakage (*19*). We, therefore, conclude that as the result of the sarcolemmal fragility developed during the preceding anoxia, cardiac beating on reoxygenation induces massive leakage of sarcoplasmic enzymes.

III. REOXYGENATION INJURY OF MITOCHONDRIA

A hasty conclusion from the results in Fig. 2 might be that formation of oxygen radicals is unlikely to be a main pathway of reperfusion injury. In fact, reoxygenation without beating induces only a small increase in release of enzyme activity into the perfusate, which is a marker of sarcolemmal disintegration. However, at this time, the mitochondrial membrane is found to be greatly impaired, probably by oxygen radicals. Figure 3 shows the composition of AST isozyme in the sarcoplasmic compartment determined by a digitonin perfusion technique (*20*). The ratio of the activity of mitochondrial AST (mAST) to that of cytoplasmic isozyme (cAST) in anoxic heart is about 2% irrespective of the duration of anoxia. When an intact heart is perfused with digitonin, a similar ratio

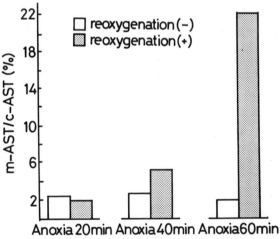

Fig. 3.   Release of mAST into the sarcoplasm upon reoxygenation. The sarcoplasmic compartment was obtained by digitonin perfusion (*20*) after the indicated period of anoxia (open column) and 10 min of subsequent reoxygenation (stippled column). The activities of AST isozymes were measured and the mAST/cAST ratio was calculated.

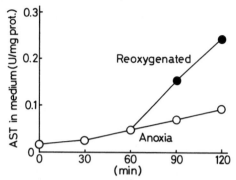

Fig. 4. Release of AST from isolated mitochondria during reoxygenation. Mitochondria isolated from rat liver were incubated under anoxic conditions for 60 min and then reoxygenated. AST activity in the medium was measured at the indicated times.

of mAST/cAST is obtained. Thus the presence of less than 2% mAST is an artifact in this technique, and there seem to be no mitochondrial enzymes in anoxic sarcoplasm. Reoxygenation after 20 min of anoxia does not change this ratio. After 40 and 60 min of anoxia, however, the ratio upon reoxygenation increases two- and tenfold, respectively. These results indicate that mitochondrial enzymes are released into the cytosol on reoxygenation, and that certain pathological changes induced by more than 40 min of anoxia are requisite for this mitochondrial injury. As described above, the electron transport system in mitochondria is one of the most probable sources of oxygen radicals in the early phase of reperfusion, and it is therefore reasonable that mitochondria are preferentially damaged on reoxygenation.

Leakage of matrix enzymes is also observed when isolated mitochondria are reoxygenated (Fig. 4) or incubated with a xanthine oxidase system which generates oxygen radicals (data not shown). We have examined the protective effects of various compounds against anoxic and reoxygenation injury of isolated mitochondria (Table I). The fact that external addition of ATP prevents both types of injury is an important clue to the exact point at which ischemic injury becomes irreversible. While sufficient extramitochondrial ATP, namely sarcoplasmic ATP, is present, mitochondria are considered to be protected against both anoxic and oxidative injury. We have shown that there are two compartments of ATP in the heart, sarcoplasmic free ATP and protein-bound, or mitochondrial ATP, and that during ischemia the free ATP level de-

TABLE I

Protection against Anoxic and Reoxygenation Injury of Mitochondria

Addition	Anoxia	Reoxygenation
ATP	+	+
EGTA	+	+
Inhibitors of PLA$_2$	+	−
Ruthenium red	−	+
Cyclosporin A	−	+
Catalase+SOD	−	−

Isolated mitochondria were incubated with various kinds of agents under anoxic conditions for 90 min (anoxia), or aerobic conditions for 60 min after 30 min of anoxic incubation without any agents (reoxygenation). Anoxic injury was assessed by the respiratory control ratio, and reoxygenation injury by the release of mAST into the medium. Plus signs indicate that injuries were markedly reduced by the addition, minus signs indicate little or no change from that on incubation without the addition. PLA$_2$, phospholipase A$_2$; SOD, superoxide dismutase.

creases more rapidly (21). The preceding anoxic change that is needed for development of reoxygenation injury of perfused heart may be depletion of sarcoplamsic ATP.

The protective effect of EGTA on reoxygenation injury indicates a Ca^{2+}-dependent pathway, as found in anoxia, but the two mechanisms of membrane disintegration are different. A transition metal chelator, which inhibits the conversion of H$_2$O$_2$ to a hydroxyl radical, protects reoxygenated mitochondria only partially, and hence EGTA still acts mainly as a Ca^{2+} chelator. However, the amount of fatty acids accumulated during anoxia is reduced by oxygenation, probably due to re-acylation, and inhibitors of phospholipase A$_2$ are ineffective, indicating that this Ca^{2+}-dependent enzyme is not involved in the reoxygenation injury. Ruthenium red, an inhibitor of the Ca^{2+} uptake system of mitochondria, prevents reoxygenation injury. Thus during respiration, Ca^{2+} appears to be taken up and to damage the membrane from the matrix side. EGTA is also suggested to inhibit the uptake by its action in chelating the ion released. Vlessis and Mela-Riker (22) demonstrated that H$_2$O$_2$-induced mitochondrial Ca^{2+} efflux leads to Ca^{2+} reuptake, establishing a futile cycle. Our data shown here indicate that the reuptake of released Ca^{2+} is the direct cause of reoxygenation injury.

Cyclosporin A is a cyclic polypeptide that is widely used as an immunosuppressive agent, and was recently reported to inhibit Ca^{2+}-induced increase in membrane permeability of isolated mitochondria (23). There are many reports that a large amount of Ca^{2+} induces an increase in non-specific permeability of the membrane (for review, see ref.

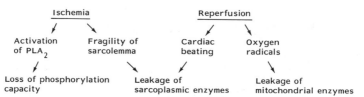

Fig. 5. Proposed mechanisms of ischemic and reperfusion injuries of the heart. PLA$_2$, phospholipase A$_2$.

24). Early reports suggested that mitochondria become leaky to small molecules, but recently they were also found to release matrix proteins (25). Pfeiffer and co-workers have reported that energy-dependent uptake of Ca^{2+} followed by addition of some inducing agents activates phospholipase A$_2$, leading to increased membrane permeability (26). In fact, inhibitors of the enzyme attenuate the permeability (27). However, cyclosporin A does not affect phospholipase A$_2$ activity (28), and so Crompton et al. (23) and Gunter and Pfeiffer (24) suggest that the inner membrane contains a Ca^{2+}-activated pore or channel that is inhibited by cyclosporin A. The release of mAST from reoxygenated mitochondria, shown in this paper, is also sensitive to cyclosporin A, and independent of phospholipase A$_2$. The actual existence of this type of pore is still uncertain, but clearly there are at least two mechanisms of disintegration of the mitochondrial membrane, and that of reoxygenation injury is quite different from that of anoxic injury.

Since the xanthine oxidase system also induces enzyme release, the formation of oxygen radicals is definitely an initial event in reoxygenation injury. However, the external addition of radical scavengers, such as superoxide dismutase and catalase, either singly or in combination, does not have any protective effect on enzyme release. This can be explained by impermeability of the mitochondrial membrane to the added enzymes. We thus consider that oxygen radicals generated in the matrix lead to the injury. Cytoplasmic or extracellular scavenging systems, therefore, cannot prevent mitochondrial injury upon reperfusion of ischemic heart. The contribution of the intramitochondrial scavenging system must be investigated.

IV. CONCLUDING REMARKS

The processes of ischemic and reperfusion injury of the heart are summarized in Fig. 5. Ischemic injury is mainly caused by increase in the cytosolic Ca^{2+} level as the result of energy insufficiency. Reperfusion injury overlaps the preceding ischemic changes and involves damage caused by resumption of cardiac beating and generation of oxygen radicals upon reoxygenation. In clinical cases, such as acute myocardial infarction with thrombolytic therapy or heart transplantation, it may not be necessary to distinguish the two processes. However, to develop effective preventive and therapeutic strategies, extensive analyses of basic molecular biological phenomena are required.

SUMMARY

Ischemia results in insufficiency of energy metabolism and alteration of cellular Ca^{2+} homeostasis. Prolonged anoxia increases the cytoplasmic Ca^{2+} concentration, resulting in several types of Ca^{2+}-dependent cell injury. Mitochondria hold the key to reversibility in ischemic organs. During anoxic incubation of isolated mitochondria, decrease in the ATP level is followed by release of matrix Ca^{2+}. We found that loss of the oxidative phosphorylation capacity is the result of activation of mitochondrial phospholipase A_2 by the released Ca^{2+}. Reperfusion of an ischemic organ is essential for its recovery, but paradoxically can also increase injury. Reoxygenation of anoxic rat heart induces massive leakage of myocardium enzymes as a result of sarcolemmal disintegration. We found that the major cause of this leakage is resumption of cardiac beating upon reoxygenation rather than oxygen radicals. In ischemic heart, oxygen radicals generated on reoxygenation seem to injure mitochondria preferentially. Reoxygenated mitochondria lose their diffusion barrier against matrix proteins. The exact mechanism of this loss is unknown, but it is also Ca^{2+}-dependent and independent of phospholipase A_2. Thus, the mechanism of reoxygenation injury of mitochondria is completely different from that of anoxic injury. Interestingly, addition of ATP protected isolated mitochondria against both anoxic and oxidative injury. Depletion of ATP in ischemic heart is, therefore, the initial and indispensable event in both ischemic and reperfusion injury.

Acknowledgments
This study was supported in part by Grants-in-Aid for Scientific Research from the Ministry of Education, Science, and Culture of Japan. We thank Ms. K. Oshita-Chihara for technical assistance.

REFERENCES

1. Watanabe, F., Kamiike, W., Nishimura, T., Hashimoto, T., and Tagawa, K. *J. Biochem.*, **94**, 493 (1983).
2. Kamiike, W., Watanabe, F., Hashimoto, T., Tagawa, K., Ikeda, Y., Nakao, K., and Kawashima, Y. *J. Biochem.* **91**, 1349 (1982).
3. Shug, A.L., Shrago, E., Bittar, N., Folts, J.D., and Koke, J.R. *Am. J. Physiol.*, **228**, 689 (1975).
4. Mittnacht, S. and Farber, J.L. *J. Biol. Chem.* **256**, 3199 (1981).
5. Smith, M.W., Collan, Y., Kahng, M.W., and Trump, B.F. *Biochim. Biophys. Acta*, **618**, 192 (1980).
6. Chien, K.R., Abrams, J., Serroni, A., Martin, J.T., and Farber, J.L. *J. Biol. Chem.*, **253**, 4809 (1978).
7. Cheung, J.Y., Bonventre, J.V., Malis, C.D., and Leaf, A. *N. Engl. J. Med.*, **314**, 1670 (1986).
8. Carafoli, E. *Annu. Rev. Biochem.*, **56**, 395 (1987).
9. Rossi, C.S. and Lehninger, A.L. *J. Biol. Chem.*, **239**, 3971 (1964).
10. Kloner, R.A., Przyklenk, K., and Whittaker, P. *Circulation*, **80**, 1115 (1989).
11. McCord, J.M. *N. Engl. J. Med.*, **312**, 159 (1985).
12. Lucchesi, B.R. *Annu. Rev. Pharmacol. Toxicol.*, **26**, 201 (1986).
13. Chance, B., Sies, H., and Boveris, A. *Physiol. Rev.*, **59**, 527 (1979).
14. Nishida, T., Inoue, T., Kamiike, W., Kawashima, Y., and Tagawa, K. *J. Biochem.*, **106**, 533 (1989).
15. Watanabe, F., Hashimoto, T., and Tagawa, K. *J. Biochem.*, **97**, 1229 (1985).
16. Nishimura, T., Yoshida, Y., Watanabe, F., Koseki, M., Nishida, T., Tagawa, K., and Kawashima, Y. *Hepatology*, **6**, 701 (1986).
17. Hearse, D.J. *J. Mol. Cell. Cardiol.*, **9**, 605 (1977).
18. Feuvray, D. and deLeiris, J. *J. Mol. Cell. Cardiol.*, **7**, 207 (1975).
19. Takami, H., Matsuda, H., Kuki, S., Nishimura, M., Kawashima, Y., Watari, H., Furuya, E., and Tagawa, K. *Pflügers Arch.*, **416**, 144 (1990).
20. Quistorff, B. and Grunnet, N. *Biochem. J.*, **226**, 289 (1985).
21. Takami, H., Furuya, E., Tagawa, K., Seo, Y., Murakami, M., Watari, H., Matsuda, H., Hirose, H., and Kawashima, Y. *J. Biochem.*, **104**, 35 (1988).
22. Vlessis, A.A. and Mela-Riker, L. *Am. J. Physiol.*, **256**, C1196 (1989).
23. Crompton, M., Ellinger, H., and Costi, A. *Biochem. J.*, **255**, 357 (1988).
24. Gunter, T.E. and Pfeiffer, D.R. *Am. J. Physiol.*, **258**, C755 (1990).
25. Igbavboa, U., Zwizinski, C.W., and Pfeiffer, D.R. *Biochem. Biophys. Res. Commun.*, **161**, 619 (1989).
26. Pfeiffer, D.R., Schmid, P.C., Beatrice, M.C., and Schmid, H.H.O. *J. Biol. Chem.*, **254**, 11485 (1979).
27. Broekemeier, K.M., Schmid, P.C., Schmid, H.H.O., and Pfeiffer, D.R. *J. Biol. Chem.*, **260**, 105 (1985).
28. Broekemeier, K.M., Dempsey, M.E., and Pfeiffer, D.R. *J. Biol. Chem.* **264**, 7826 (1989).

Mitochondrial Disease and the Ageing Process: From Microbe to Man

ANTHONY W. LINNANE AND PHILLIP NAGLEY

Department of Biochemistry and Centre for Molecular Biology and Medicine, Monash University, Clayton, Victoria 3168, Australia

The biogenesis of mitochondria has long been an interest of our laboratory (*1, 2*). Until recently such studies have concentrated on the formation of mitochondria in yeast cells, because of the great advantages for genetic and physiological manipulation offered by this simple eukaryotic organism (particularly baker's yeast *Saccharomyces cerevisiae*). One may ask what is the relevance of yeast mitochondria and their molecular biology and bioenergetics to a symposium on the molecular biology of the myocardium. It is the intention of this article to indicate the relevance of yeast mitochondrial biogenesis to that of human cells, and thereby to energy production in the myocardium and in cardiac pathology. Indeed, there are extensive parallels between yeast and mammalian cell systems in terms of mitochondrial biogenesis and cellular bioenergetics based on mitochondrial function. A further important generalisation that will be developed here is that mitochondrial myopathies are but one element of a much broader picture in regard to disease and the ageing process.

I. GENERAL PRINCIPLES OF MITOCHONDRIAL BIOGENESIS

The outstanding feature of the biogenesis of mitochondria is the cooperative interaction of two genetic systems in eukaryotic cells (*1-6*). The major chromosomal system encodes many hundreds of mitochondrial proteins that are biosynthesised on cytosolic ribosomes and imported into mitochondria, to be distributed by a sophisticated targeting system to the appropriate subcompartment of the organelle. A handful of proteins are encoded inside mitochondria by mitochondrial DNA (mtDNA), and are synthesized on mitochondrial ribosomes using mitochondrial mRNA as a template for translation. These ribosomes contain rRNA encoded by mtDNA, and interact with a set of mitochondrially encoded tRNA molecules. In general, mitochondrially encoded proteins constitute a set of integral membrane subunits that assemble with nuclearly encoded imported proteins to produce some of the key enzyme complexes of oxidative phosphorylation. These include the respiratory complexes NADH-CoQ reductase (Complex I), cytochrome bc_1 (Complex III), cytochrome c oxidase (Complex IV) and the proton-translocating ATP synthase (Complex V).

Mitochondrial structures in cells probably arose at the dawn of the eukaryotic age through an early intracellular symbiosis between an invading bacterium-like entity and a host cell that developed a nuclear membrane and chromosomes (7). The modern mitochondria that were shaped during the course of subsequent evolution retain only a dozen or so key genes within the organellar membranes. A semi-autonomous gene expression system produces mitochondrial proteins endogenously. Most of the genes of the early invaders have been transferred to the nucleus during hundreds of million of years of evolution, with many of the proteins having acquired N-terminal cleavable leader sequences to target the passenger moiety of precursors into mitochondria where the leader is proteolytically cleaved in the matrix compartment (8). These principles are the same for both yeast and mammals. In the comparative surveys of mitochondrial molecular biology and bioenergetic physiology in yeast and mammals that follow, there is to be found the overriding theme of unity in both biochemical principle and molecular genetic phenomenon. This unitary theme transcends the specific differential detail of gene organization or expression and the manifestation of mitochondrial mutation in the disparate cellular contexts. In general, the prior detailed studies on yeast have provided mechanistic and conceptual frameworks for present insight into human systems in health and disease.

II. MITOCHONDRIAL GENOME ORGANISATION AND EXPRESSION

Salient features of the mitochondrial genomes of yeast and humans are summarized in Table I. Whilst the mtDNA genome of *S. cerevisiae* is about five times larger than that of humans, the number of genes in

TABLE I

Mitochondrial Genomes of Yeast and Humans and Their Major Coding Functions

Feature	Yeast	Human
mtDNA genome size (kb)	70–85	16.6
Number of mtDNA genomes per cell	50–100	1,000–5,000
Ribosomal RNA genes	15S, 21S	12S, 16S
Transfer RNA genes	25 tRNAs	22 tRNAs
Protein subunits encoded:		
ATP synthase	3	2
Cytochrome b	1	1
Cytochrome c oxidase	3	3
NADH-CoQ reductase	—	7
Ribosomal protein	1	—
Total proteins	8	13
Genome organization	Extended intergenic regions (A, T-rich); tRNA genes clustered	Little or no space between genes; tRNA gene punctuation
Introns	In some genes	No introns

Detailed citations of references are found in refs. *4–6*.

terms of proteins encoded is actually smaller. The yeast mitochondrial genome is characterised by extended A,T-rich spacer regions between genes, as well as introns in several genes, in contrast to the highly compact human mtDNA genome within which spacers between genes are the exception rather than the rule. Indeed, the *S. cerevisiae* mtDNA genome is rather plastic in the sense that in different strains certain introns may or may not be present (so-called optional introns) and the intergenic spacer regions themselves vary in size and sequence. On the other hand, the compact mtDNA genome of humans typifies the tight organisation found in many other mammals and, indeed, in vertebrates in general.

Nonetheless, sequence variation between individual humans occurs to the extent that restriction site polymorphisms (as a convenient indicator of sequence divergence) provide very useful markers in a range of genetic and anthropological investigations. For example, mtDNA is readily shown to be maternally inherited (*6*). Detailed studies of population groups have been made, even to the point of ascribing possible ancestry of modern humans to a small progenitor group within inferred geographical limits (*9*). Yet the sequence variability amongst humans presents problems in the rapid determination of the relevant mutation responsible for a particular maternally inherited mitochondrial disease phenotype (see below). It has been necessary to carry out extensive comparative sequence analysis of large tracts of the mtDNA genomes of a range of patients and unaffected controls (see *e.g.* refs. *10–12*).

In terms of genome dosage, in yeast the hundred or so mtDNA genomes account for about 10–20 percent of total cellular DNA, due to the very small nuclear genome size of *S. cerevisiae*. In humans, in the context of the nuclear genome being orders of magnitude larger than that of yeast, the mtDNA genome accounts for less than 1% of cellular DNA.

Yet the coding functions of both sorts of mitochondrial genomes (Table I) are remarkably similar: a pair of rRNA genes; a panel of tRNA genes adequate for translation of the specialised genetic codes in mitochondria (different in yeast and humans, and each distinct from that of the nucleo-cytosolic system); and a set of protein-coding genes. The main differences in the proteins encoded are as follows. The two mitochondrial ATP synthase genes in humans encode hydrophobic proteins of the proton channel F_0-sector of the complex. A further key F_0 component, subunit 9, is nuclear encoded in mammals but has remained a mitochondrial gene product in yeast. Human mtDNA encodes seven subunits of the NADH-CoQ reductase enzyme complex (genes known generically as ND). *S. cerevisiae* lacks any mitochondrial genes encoding subunits of this complex, probably because the enzyme corresponding to complex I in yeast is simpler than that in humans, lacking proton pumping capacity. All mitochondrial ribosomal proteins in humans, belonging to both the small and large ribosome subunits, are nuclear encoded. This generality applies to all proteins of ribosomal subunits in yeast mitochondria, with the exception of a single mitochondrially encoded protein of the small subunit of the ribosome (known as var1, because of its variable size that is dependent upon optional in-frame insertions in the coding region of different strains).

Transcription of human mtDNA (*5*) takes place from two divergent promoters located in the D-loop region of mtDNA (a key region involved in the initiation of mtDNA replication). Each strand of DNA is transcribed to completion to generate a single precursor for all genes in each of the respective constituent strands of mtDNA. Post-transcriptional processing of these RNA precursors, probably flagged by the tRNA sequences that lie between many of the other genes, generates individual gene-specific transcripts. In mtDNA of *S. cerevisiae*, a small number of promoters generate very long precursor transcripts encompassing several genes that are subsequently cleaved to provide monocistronic messenger RNAs (in some cases, polycistronic), tRNAs and rRNAs. Transcripts of genes containing introns must be spliced. This has provided a popular area of molecular biological research on yeast mitochondria since some introns contain their own coding regions for proteins enhancing the splicing reaction (maturases) while other precursor molecules undergo self-splicing (ribozyme-type catalysis). These particularities of mitochondria in yeast (*4, 5*) nevertheless should not obscure the major principles of mitochondrial gene expression. Its purpose in all eukaryotes is to drive the biosynthesis of protein subunits that assemble on the inner mitochondrial membrane, thus representing the essential contributions of mtDNA to the energy-transducing respiratory enzymes and ATP synthase complexes that sustain aerobic cellular performance in mammals and in yeast.

III. MUTATIONS OF YEAST AND HUMAN MITOCHONDRIAL GENOMES

Historically, one of the primary discoveries that led to the identification of the mitochondrial genome, as such, was the recognition of non-Mendelian or cytoplasmic inheritance in microorganisms such as yeast. This led directly to the development of the field of mitochondrial genetics. With the elucidation in the mid 1970's of the first physical map of genes in the yeast mitochondrial genome (*13*), the molecular genetics of mitochondria had indeed come of age. Key mutants in this field of yeast mitochondrial genetics (Table II) were drug-resistant mutants affecting mitochondrial ribosomes (CAP^R or ERY^R) or enzyme complexes such as ATP synthase (OLI^R) or complex III (ANA^R), *mit*$^-$ mutants that contain small lesions in mtDNA affecting particular subunits of respiratory enzyme complexes or ATP synthase, or the cytoplasmic petite mutants that harbor extensive deletions of mtDNA sequences (*rho*$^-$ petites), in some cases leading to total deletion of detectable mtDNA (*rho*^{0}).

Collectively these mutants played key roles in assigning gene-product relationships through biochemical genetic analysis of individual mutant strains (drug resistant or *mit*$^-$). Physical mapping studies depended on the determination of the genetic content of the residual DNA in individual *rho*$^-$ petites, in terms of the drug resistance or *mit*$^-$ genetic loci retained or deleted, and the physically defined segment of mtDNA retained in each such *rho*$^-$ petite. Libraries of *rho*$^-$ petites were established which collectively covered the entire yeast mtDNA genome (*3*). Piecing together this molecular jigsaw of deletions not only allowed physical

TABLE II
Congruent Effects of Mutation and Inhibitors on Yeast and Human Mitochondria

Feature	Yeast	Human
Mutations in mtDNA		
Drug resistance	*e.g.* CAP^R, ERY^R	*e.g.* CAP^R
	OLI^R ANA^R	OLI^R ANA^R
Small lesions in:		
protein genes	*mit$^-$*	LHON
RNA genes	*syn$^-$* (*e.g.* tRNA)	MERRF (*e.g.* tRNA)
Large deletions	*rho$^-$*	KSS; CPEO
Total mtDNA deletion	*rho^0*	*rho^0*
Inhibitors of mitochondrial	*e.g.* CAP, ERY,	*e.g.* CAP,
development	EthBr	EthBr, AZT
Anaerobic growth of cells:	Yes	Yes
	(facultative anaerobic)	(*plus* pyruvate)
Cytochromes depleted	Yes	Yes
NADH reoxidation	Acetaldehyde	Pyruvate
shuttle	to ethanol	to lactate

Detailed citations of references are made as follows: mitochondrial genetics in yeast (*3, 15*) and humans (*6*); human diseases (*10, 21, 23*); action of inhibitors of mitochondrial bio-genesis (*1–6*). Other aspects including anaerobic growth and the effect of AZT are discussed in the text (see also Table III). Abbreviations of drugs: CAP, chloramphenicol; ERY, erythromycin; OLI, oligomycin; ANA, antimycin A; EthBr, ethidium bromide; AZT, zidovudine. Abbreviations of clinical syndromes: LHON, Leber's hereditary optic neur-opathy; MERRF, myoclonic epilepsy and ragged red fibre disease; KSS, Kearns-Sayre syndrome; CPEO, chronic progressive external ophthalmoplegia.

mapping of genetic markers used to characterise the petite mtDNA genomes in the first place, but also enabled new mutations to be physi-cally mapped within a few days (*14*). The picture that we had concerning the physical map of genes in yeast mtDNA was subsequently filled in by DNA sequence data on the protein coding regions, and rRNA and tRNA genes (*4, 15*). The determination of this picture in yeast (*16*) was essential to the interpretation of the nucleotide sequences that emerged from the total sequencing of human mtDNA in Sanger's laboratory (*17*). Assign-ment of ND reading frames had to wait a number of years (*18*) since yeast mtDNA has no such genes (see above). Transfer RNAs and ribosomal RNAs were identified by reference to cognate functional RNAs from a range of genomes across the biological world.

As the phenomenology of mitochondrial mutations emerges in the study of cultured mammalian cells on the one hand, and human patients on the other, the prior information and the concepts that have been shaped through intensive analysis of yeast provide the framework for understanding the mammalian systems. Thus, the analyses of drug resistant mutants in cultured mammalian (including human) cell lines revealed mutants such as CAP[R] and OLI[R] formally analogous to those previously described in yeast. The behaviour of such mutations, analysed by somatic cell genetic techniques in mammalian cell lines, revealed features of non-Mendelian genetics amongst populations of cytoplasmic mtDNA molecules (6), exactly as had been realised for the first time some years before by study of yeast petite mutants and later drug-resistant mutants in yeast (2, 13). Whilst the lack of recombination between mitochondrial genetic markers in mammalian cells (6) distinguished their phenomonology from that of the highly recombinational mitochondrial genomes of yeast (3, 15), the broad principles of heteroplasmy, cytoplasmic segregation and nuclear modulation of mitochondrial genetic expression were clearly mirrored in both microbe and man.

Recent discoveries of the molecular basis of several mitochondrial diseases in the past few years emphasise the parallels between human and yeast mitochondrial genetics. Myopathies and neuropathies that have a recognisable degree of maternal inheritance have focussed interest on the possibility of specific mitochondrial mutations being responsible for the clinical symptoms. This turned out to be the case for the clearly maternally inherited diseases Leber's Hereditary Optic Neuropathy (LHON), in which a mutation in the ND4 gene (determining the substitution Arg →His) was recognised (19); this would be classified as a *mit⁻* mutation in yeast. More recently, a human mutation associated with both myopathy and neuropathy was shown to be equivalent to a *mit⁻* mutation in ATP synthase subunit 6 (12). Certain patients with myoclonic epilepsy ragged red fibre disease (MERRF), also affected by a maternally inherited genetic defect, have been recently shown to contain a mitochondrial tRNA[Lys] with a base change (20). As this mutation probably affects protein synthesis in mitochondria, it is of the class of *syn⁻* mutations recognised in yeast many years ago in which RNA molecules of the mitochondrial protein synthesising system of yeast are primarily affected (4).

Perhaps the most widely studied mitochondrial diseases from the aspect of mtDNA characterisation have been Kearns-Sayre syndrome (KSS) or chronic progressive external opthalmoplegia (CPEO) which lead to chronic muscle fatigue or neuromuscular dysfunction (21, 22). These conditions, that do not necessarily show clear maternal inheritance pedigrees, are characterised by the accumulation of subgenomic mtDNA molecules in coexistence with normal full length mtDNA genomes in affected cells and tissues. The analogy with *rho⁻* petite genomes is obvious from a molecular genetic perspective, yet the physiological consequences are more complex in humans than in yeast, because the viability and robustness of tissues in the intact mammalian organism do generally depend on the presence and proper expression of intact mtDNA molecules. By contrast in yeast, a facultative anaerobe, cells containing exclusively the deleted subgenomic mtDNA molecules are viable as long as cells grow on fermentable substrates such as glucose. This tenuous distinction between yeast and man has now dramatically fallen by the wayside, certainly for tissue culture cells and probably for intact human tissues too (see below). Avian and mammalian cells totally devoid of mtDNA (*rho⁰*) have been generated, as for yeast, together with simple conditions for maintaining cell viability under strictly fermentative conditions that dispense with aerobic respiratory metabolism (as amplified later in the next section).

The formation of products of mitochondrial protein synthesis can be blocked in both yeast and human cells by use of bacterial-type ribosome inhibitors such as chloramphenicol and erythromycin. Inhibition of mitochondrial development at the level of ablation of the mitochondrial genome itself has been achieved in both yeast and mammalian cells using the DNA-binding drug ethidium bromide. In such eukaryotes, representing the extremes of cellular complexity, this drug is an effective inducer of the *rho⁰* state, probably due to its potent inhibition of mtDNA replication (but not that of nuclear chromosomes). The technical manoeuvre of including pyruvate and uridine with ethidium-treated mammalian cells provides essential support for the non-mitochondrially dependent growth pattern (see next section). The *rho⁰* tissue culture cells represent the ultimate in mtDNA deletion in a continuum evident in the patients carrying partially deleted mtDNA molecules (see above). The recently characterised properties of anaerobic growth of human cells reinforces the parallel behaviour of mammalian and yeast cells as will now be discussed.

IV.   CELLULAR GROWTH WITHOUT MITOCHONDRIAL FUNCTIONS

Baker's yeast *S. cerevisiae* is well known as a facultative anaerobe. It can grow aerobically on non-fermentable substrates such as ethanol, glycerol or lactate, where it uses mitochondrial oxidative phosphorylation to reoxidise NADH and to generate ATP. Carbon sources are oxidised completely to CO_2. Such yeast cells can grow alternatively on glucose. In this case they utilise the glycolytic pathway to generate pyruvate initially. Two moles of ATP are produced by substrate level phosphorylation and two moles of NADH results from each mole of glucose fermented. The critical task faced by the cell is to dispose of excess reducing power: NADH must be reoxidised. In fermenting yeast, this is achieved through conversion of pyruvate to acetaldehyde (by pyruvate decarboxylase) and then reduction of acetaldehyde to yield ethanol, a reaction catalysed by alcohol dehydrogenase that simultaneously regenerates NAD^+ from NADH. In yeast, the presence of glucose has a catabolite repression effect on the development of functional mitochondrial respiratory system, so mitochondria are functionally less active during fermentative growth. A similar pattern of bioenergetics in terms of fermentation, ATP production and NADH regeneration occurs during anaerobic growth. In this condition, however, the typical mitochondrial cytochromes a, a_3, b, c_1, c are not made, because oxygen is required for haem biosynthesis.

How do human cells behave in these terms? They can carry out respiratory growth like yeast using non-fermentable substrates; significantly, this also occurs on glucose substrates, since the catabolite repression phenomenon does not generally apply in mammalian cells. Thus in aerobic cells a glucose substrate is fully oxidised to CO_2, the NADH being regenerated by electron transport on the mitochondrial respiratory chain using oxygen as terminal electron acceptor. Nevertheless, human muscle cells working hard can in steady state be physiologically similar to fermenting yeast. They glycolyse glucose so fast to generate ATP by substrate level phosphorylation for muscle contraction, that the mitochondria (probably also partially deficient in oxygen) are unable to keep up. NADH is removed by lactate dehydrogenase that catalyses reduction of pyruvate to lactate. The muscles thus go into short-term oxygen debt; lactate is circulated in the blood to the liver, for reoxidation to pyruvate.

The discovery that avian (24) and mammalian cells (25) could be treated with ethidium bromide to become rho^0 (totally deficient in mtDNA) indicates that indeed viable cells can be formed that do not have a functional oxidative phosphorylation system in mitochondria. The growth of such rho^0 vertebrate cells is possible only if both pyruvate and uridine are included in the growth media (24). Pyruvate, in principle, is required as substrate for lactate dehydrogenase to enable the reoxidation of NADH. It is not clear why excess pyruvate is needed, since the stoichiometry of glucose fermentation should yield equivalent amounts of pyruvate and NADH; nevertheless; on empirical grounds additional pyruvate is clearly needed. Uridine must be present in the growth medium, since its biosynthesis requires a functional mitochondrial respiratory system to oxidize substrates in the pyrimidine biosynthetic pathway.

These data have been recently extended in two directions by work in our laboratory. First, AZT (zidovudine) has been reported by other laboratories to cause mitochondrial myopathies in patients undergoing prolonged treatments with this drug (26). We have recently shown (27) that cultured human cells (Namalwa) will grow aerobically in the presence of AZT only when pyruvate is added (Table III). Thus AZT strongly compromises mitochondrial bioenergetic functions, and this probably takes place as a result of inhibition of mitochondrial transcription of output due to the powerful adverse effects of AZT on mitochondrial DNA replication (28). Second, we have posed the question: can human cells be made to grow anaerobically? The answer is clearly yes, but only as long as pyruvate is included in the growth medium (Table III). A strict requirement for uridine remains to be established. So here

TABLE III

Growth of Human Namalwa Cells in the Presence of Inhibitors of mtDNA Replication or under Anaerobic Conditions

Conditions of culture			Doubling time (hr)	Glucose consumption (nmol per 10^5 cells during doubling)
Atmosphere	Drug	Pyruvate		
Air	—	—	20	410
Air	EthBr	—	Death	N.D.
Air	EthBr	+	48	1,540
Air	AZT	—	Death	N.D.
Air	AZT	+	48	N.D.
N_2	—	—	Death	0.0
N_2	—	+	20	920

Namalwa cells (a human lymphoblastoid cell line) were grown in 20 ml suspension cultures, inoculated at 10^5 cells/ml, in RPMI-1640 medium containing 50 μM 2-mercaptoethanol, 10 mM HEPES buffer, 10 IU/ml penicillin and 10 μg/ml streptomycin, supplemented with 10% fetal calf serum. The atmosphere was 5% CO_2 with the balance as indicated. Other additions to the medium: EthBr (50 ng/ml), AZT (400 μM), pyruvate (1 mM). Cells treated with EthBr or AZT had been precultured for at least 10 generations in the respective drug in the presence of pyruvate. Cells for anaerobic growth were transferred directly from an untreated aerobic culture. Cells were then monitored over a four-day growth period. Glucose consumption was determined over a 24 hr period using a glucose analyser, and calculated using a formula yielding an adjusted average per 10^5 cells. "Death" indicates cells died after 1–2 days of culture. N.D. indicates not determined. Data from F. Vaillant, B.E. Loveland, P. Nagley, and A.W. Linnane (ref. 27 and unpublished).

again is a direct parallel to yeast cells. Mammalian cells in culture are indeed facultative anaerobes. These anaerobic cells (Namalwa) can be cultured for prolonged periods in the absence of oxygen (but with pyruvate). The mitochondrial cytochromes disappear just as in yeast, and the respiratory system therefore loses its active enzyme complexes. These cytochromes and respiratory enzymes reappear after a period of aeration of the previously anaerobic cells (27). Note the increased glucose consumption (Table III) of EthBr-treated and anaerobically grown cells, indicative of their reliance on glycolytic fermentation to produce ATP in the absence of oxidative phosphorylation.

The significance of all this is plainly that mammalian cells must be thought of as bioenergetically resilient in the sense that when pressed very hard under laboratory conditions mitochondrial function is indeed ultimately dispensable. How does this relate to the human organism? Surely it forces us to think of our cells and tissues in more subtle ways than merely considering mitochondria as a *sine qua non* of cellular viability and function. The remaining section of this article will draw on these notions to present our current views on the major question of the mitochondrial genetic contribution to disease and ageing. These considerations provide a backdrop to the more specific instances of mitochondrial diseases that include cardiomyopathies and muscular energy deprivation due to inherited or somatic lesions in mtDNA.

V. MITOCHONDRIAL GENE MUTATION: AGEING AND DISEASE

As discussed above, mitochondrial DNA mutation is prevalent in human cells. Because of the role of mtDNA in contributing essential components of the respiratory system and ATP synthase, such mutations would be expected to lead to a debilitation in the mitochondrially localised production of cellular energy. Indeed, it has been proposed that such mutations accumulate in the mtDNA of somatic cells of ageing humans right from an early age, and further, that a major contribution to the loss of cellular bioenergetic output in ageing humans is due to this mitochondrial accumulation of mutation (29). In support of this view, there is an age-related decline in the bioenergetic function of skeletal muscle (30). Significantly, in sections of heart muscle from patients of different ages, histochemical analysis revealed a mosaic of cells in which cells deficient in cytochrome *c* oxidase were found in clusters adjacent to cells with a normal cytochrome *c* oxidase histochemical response (31). This can be readily ascribed to a somatic segregation of a mutant mtDNA molecule, having arisen possibly many cell divisions before, perhaps even as early as the first stages of embryogenesis. This segregation, most likely from a heteroplasmic progenitor cell which contained a mixture of wild-type and mutant mtDNA genomes, is one of the hallmarks of extranuclear inheritance, especially where a high multiplicity in copy number of the organelle genome occurs, as in yeast or humans. To understand this situation in more detail, it is useful to consider our recent results on the accumulation of a particular change in mtDNA in humans of different ages (32). As mentioned above, a number of myopathies are characterised by the presence of subgenomic deleted molecules of mtDNA coexisting with full length mtDNA. We have carried out a study of ten human subjects, ranging from 80 min to 87 years, on a particular 5 kb deletion that is representative of one of the more frequently occurring deletions first identified (33) in diseased tissues, including cardiomyopathies (34), and summarised by Grossman (23). This particular deletion represents a "hot-spot" for excision of a subgenomic mtDNA segment (21, 35) and was readily detected by the polymerase chain reaction (PCR) technique in our study (32) on a wide range of tissue samples from eight adult patients ranging from 40 years and upwards at death. By contrast, in similar analyses on tissue samples from two infants (80 min and 3 months old) the deletion could not be detected after the conventional 30 cycles of PCR amplification. Yet after 60 cycles of PCR amplification of the infant DNA samples, the deletion could be detected. None of the patients in the study had a known mitochondrial disease. It was proposed (32) that this deletion progressively accumulates throughout life and that the occurrence of this deletion in individual tissues may contribute to the inability of the tissue to function normally. It is of course not possible to exclude that the specific deletion whose age-related abundance is characterised in this work is the only mtDNA change associated with the ageing phenomenon. Almost certainly there are other mtDNA changes (29) and it may be that these are equally or more significant in ageing, and perhaps also in mitochondrial diseases (cf. refs. 23 and 36).

Whilst it is clear that there is a correlation between mitochondrial dysfunction in pathological situations (such as in KSS or CPEO) and the presence of deleted mtDNA molecules in cells (37, 38), it has been difficult to correlate the nature or abundance of large-scale deletions with specific biochemical phenotypes or clinical manifestations. At the molecular biological level, Nakase et al. (39) suggest that subgenomic mtDNA molecules, that are transcribed but not translated, eventually become segregated into mitochondria lacking full length mtDNA. Shoubridge *et al.* (*38*) argue that the subgenomic mtDNA molecules (irrespective of the specific deletion end points) interfere with expression of normal mtDNA genomes co-existing in the same cell, possible by generating imbalance between mRNA and tRNA, leading to stalling of translation of mitochondrially encoded proteins.

Where mtDNA deletions are concerned, it is thus not always easy to reach definitive conclusions on the molecular aetiology of defined pathologies. Ozawa and colleagues (*40, 41*) report the detection of a particular 5 kb deletion in mtDNA in the residual substantia nigra tissue recovered from *post mortem* brain sections of Parkinson's disease patients (this deletion is the same one as analyzed by Linnane *et al.* (*32*)). Comparison of the frequencies of the deleted mtDNA molecules in Parkinson's patients and a young unaffected individual who died in a traffic accident indicated a 10-fold difference in the percentage of detected mtDNA molecules between affected patients (3%) and the unaffected individual (0.3%). Whilst it is possible that the extent of accumulation of the detected molecules may be directly related to Parkinson's disease (*41*), alternative interpretations are possible in view of the prevalence of this deletion in many tissues of individuals with no evident symptoms of Parkinson's disease (*32*). Either the 5 kb deletion is not directly related to the disease or there is a critical level to which the subgenomic molecules must accumulate for tissue pathology to be expressed. Moreover, it may be that this particular deletion is only one of a number of mitochondrial or nuclear gene changes that are cumulatively required for expression of the disease. The presence of deleted mtDNA molecules may not even be a primary indicator of specific disease, but probably reflects the ageing process.

Similar considerations apply to the aetiology of some cardiomyopathies. Ozawa *et al.* (*34*) presented analyses on heart tissue samples derived from cardiomyopathy patients in which subgenomic mtDNA molecules were found, the deletions being up to 7.5 kb in length. Several other suggestions have been made, in addition to mtDNA deletion, for genetic determinants involved in cardiomyopathy. For example, a gene located on chromosome 14 has been postulated to be responsible for familial hypertrophic cardiomyopathy (*42*), while an X-linked gene is also implicated in this disease (*43*). It is likely that there are multiple genetic factors contributing to the expression of cardiomyopathy.

We are clearly not yet in a position to account for the tissue-specific expression of mitochondrial disease, namely pathological conditions under which mutations in mtDNA contribute to defective mitochondrial bioenergetic function. Thus, it may be asked why in Leber's disease (LHON) the optic nerve is the earliest pathological target, when all cells in the body contain the same mtDNA mutation (cells being homoplasmic in this case). Subgenomic mtDNA molecules are also found in many tissues of KSS patients, and not only those with pronounced biochemical defects (*44*). The answers to tissue-specificity of mitochondrial disease should come from consideration of the principles above expounded in this article. Different cell types will have emphasis in their metabolic pathways in terms of balance between respiratory bioenergetic performance in mitochondria and non-mitochondrial energy production (by glycolysis). Cells have both different energy demands and the ability to perform their functions under various conditions of physiological stress. In terms of the dependence on mitochondrial functions, some cells will be less resilient in adopting the facultative anaerobic cell physiology that is clearly seen in our cultured cell studies (*27*), where the critical factor is not merely ATP production but also NADH reoxidation. Thus for a given genetic change in mtDNA, one cell type will eventually present itself as the most vulnerable in these bioenergetic terms, and its function will fail first.

Note that a broad modulation of the threshold, at which a potentially deleterious mutation in mtDNA is actually expressed as a biochemical defect, could occur even in one such cell type. This modulation depends on the accumulation of other mutations in mtDNA due to the ageing process. These mutations may individually contribute to a specific diminution in mitochondrial bioenergetic function or they may be interactive or synergistic in the amplification of defective functions. Thus a particular mutation may not in itself lead to aberrant cell physiology, but may lower the threshold at which other mutations, perhaps inherited or perhaps acquired somatically, may manifest their pathological effects. Nuclear genetic factors, such as those involved in the rate or fidelity of mtDNA replication, or in the delivery and function of nuclearly coded subunits of mitochondrial respiratory chain complexes, may also contribute to the pathology threshold for expression of mtDNA mutations. This may explain the complicated (and often obscure) familial genetics of many mitochondrial myopathies and neuropathies. The diseases also progress. This could reflect failure of more cell types due to general physiological debilitation, or perhaps to the accumulation of further mutations.

In the field of mitochondrial disease in general, and cardiomyopathy in particular, the stage is now set for detailed molecular studies of mtDNA mutation in various cell types of patients of different ages (possibly with clinically defined myopathies or neuropathies) to identify the relevant causative factors and to understand secondary effects. To do this effectively, regard must be given to the plasticity of the anerobic growth properties of many cells in terms of ATP production and NADH reoxidation. The challenge is great, but understanding the subtleties addressed here provides a resourceful approach that will enable identification of the relevant molecular and genetic factors in mitochondrially determined cardiomyopathies. One must be able to superimpose properly the pathological features due to specific mtDNA defects with the ongoing process of ageing that in itself probably leads to accumulation of sequence changes in mtDNA of most, if not all, human cells. The ageing process is in itself a degenerative disease, with mtDNA mutations contributing one additional feature, namely reduction in cellular energy production, to the other genetic and physical factors that cumulatively act in the senescence of human individuals.

SUMMARY

Crucial to the understanding of the function of the myocardium in health and disease is the molecular biology of mitochondria in which much of the cellular energy production occurs. The general principles of mitochondrial biogenesis are equivalent in human and yeast cells. The molecular genetics of mitochondria have been studied intensively for several decades in yeast, a eukaryotic microorganism amenable to ready genetic and physiological manipulation. These detailed investigations have provided the mechanistic and conceptual framework for more recent studies of mitochondrial genetic defects in human cells. Base substitutions, partial deletions and even the total loss of mtDNA have been recognised, all of which have precedents in well characterised yeast mutants. The ability of human cells to grow anaerobically under laboratory culture conditions reinforces the parallel behaviour of mammalian and yeast cells in terms of their capacities as facultative anaerobes. Cellular energy (ATP) can be obtained by anaerobic glycolysis, with the reoxidation of NADH a key function driven by pyruvate in human cells.

Against this background are considered recent data on the accumulation of mutations in mtDNA during the ageing process in humans, as well as the particular mtDNA lesions associated with more specific clinical syndromes, mainly myopathies and neuropathies. Yet biochemical and pathological changes do not always correlate well with particular mtDNA mutations and the basis of tissue-specificity of functional defects remains unclear. A general perspective is presented here, in which the pathological manifestation of a particular mitochondrial genetic change in a given tissue or cell type depends on the collateral accumulation of other mutagenic damage to mtDNA, and probably also nuclear DNA. A further critical factor in such pathology is the ability of that tissue to function with a reduced contribution from mitochondrial respiratory activity to both energy production and reoxidation of NADH. Mitochondrial disease, including certain cardiomyopathies, thus represents part of a broader picture in terms of ageing and cellular bioenergetics.

REFERENCES

1. Linnane, A.W. and Haslam, J.M. *In* "Current Topics in Cellular Regulation," Vol. 2, ed. B.L. Horecker and E.R. Stadtman, p. 101 (1970). Academic Press, New York.
2. Linnane, A.W., Haslam, J.M., Lukins, H.B., and Nagley, P. *Annu. Rev. Microbiol.*, **26**, 163 (1972).
3. Nagley, P., Sriprakash, K.S., and Linnane, A.W. *Adv. Microb. Physiol.*, **16**, 157 (1977).
4. Tzagoloff, A. and Myers, A.M. *Annu. Rev. Biochem.*, **55**, 249 (1986).
5. Attardi, G. and Schatz, G. *Annu. Rev. Cell Biol.*, **4**, 289 (1988).
6. Wallace, D.C. *In* "Birth Defects: Original Article Series," Vol. 23, p. 137 (1987). March of Dimes Birth Defects Foundation, New York.
7. Gray, M.W. and Doolittle, W.F. *Microbiol. Rev.*, **46**, 1 (1982).
8. Hartl, F.-U. and Neupert, W. *Science*, **247**, 930 (1990).
9. Cann, R.L., Stoneking, M., and Wilson, A.C. *Nature*, **325**, 31 (1987).
10. Wallace, D.C. *Trends Genet.*, **5**, 9 (1989).
11. Marzuki, S., Noer, A.S., Letrit, P., Uthanaphol, P., Thyagarajan, D., Kapsa, R., Sudoyo, H., and Byrne, E. *In* "Progress in Neuropathology," Vol. 7, ed. T. Sato and S. DiMauro, p. 181 (1991). Raven Press, New York.
12. Holt, I.J., Harding, A.E., Petty, R.K.H., and Morgan-Hughes, J.A. *Am. J. Hum. Genet.*, **46**, 428 (1990).
13. Linnane, A.W. and Nagley, P. *Plasmid*, **1**, 324 (1978).
14. Linnane, A.W., Lukins, H.B., Molloy, P.L., Nagley, P., Rytka, J., Sriprakash, K.S., and Trembath, M.K. *Proc. Natl. Acad. Sci. U.S.A.*, **73**, 2082 (1976).
15. Dujon, B. *In* "Molecular Biology of the Yeast Saccharomyces: Life Cycle and Inheritance," ed. J.N. Strathern, E.W. Jones, and J.R. Broach, p. 505 (1981). Cold Spring Harbor Laboratory, New York.
16. Borst, P. and Grivell, L.A. *Nature*, **290**, 443 (1981).

17. Anderson, S., Bankier, A.T., Barrell, B.G., DeBruijn, M.H.L., Coulson, A.R., Drouin, J., Eperon, I.C., Nierlich, D.P., Roe, B.A., Sanger, F., Schreier, P.H., Smith, A.J.H., Staden, R., and Young, I.G. *Nature*, **290**, 457 (1981).
18. Chomyn, A., Mariottini, P., Cleeter, M.W.J., Ragan, C.I., Matsuno-Yagi, A., Hatch, Y., Doolittle, R.F., and Attardi, G. *Nature*, **314**, 592 (1985).
19. Singh, G., Lott, M.T., and Wallace, D.C. *N. Engl. J. Med.*, **320**, 1300 (1989).
20. Shoffner, J.M., Lott, M.T., Lezza, A.M.S., Seibel, P., Ballinger, S.W., and Wallace, D.C. *Cell*, **61**, 931 (1990).
21. Shoffner, J.M. and Wallace, D.C. *Adv. Hum. Genet.*, **19**, 267 (1990).
22. Zeviani, M., Bonilla, E., DeVivo, D.C., and DiMauro, S. *Neurol. Clin.*, **7**, 123 (1989).
23. Grossman, L.I. *Am. J. Hum. Genet.*, **46**, 415 (1990).
24. Desjardins, P., Frost, E., and Morais, R. *Mol. Cell Biol.* **5**, 1163 (1985).
25. King, M.P. and Attardi, G. *Science*, **246**, 500 (1989).
26. Dalakas, M., Illa, I., Pekeshkpour, G.M., Laukaitis, J.P., Cohen, B., and Griffin, J.L. *N. Engl. J. Med.*, **322**, 1098 (1990).
27. Vaillant, F., Loveland, B.E., Nagley, P., and Linnane, A.W. *Biochem. Int.*, **23**, 571 (1991).
28. Simpson, M.V., Chin, C.D., Keilbaugh, S.A., Lin, T.S., and Prusoff, W.M. *Biochem. Pharmacol.*, **38**, 1033 (1989).
29. Linnane, A.W., Marzuki, S., Ozawa, T., and Tanaka, M. *Lancet*, **i**, 642 (1989).
30. Trounce, J., Byrne, E., and Marzuki, S., *Lancet*, **i**, 637 (1989).
31. Müller-Höcker, J.M. *Am. J. Pathol.*, **134**, 1167 (1989).
32. Linnane, A.W., Baumer, A., Maxwell, R.J., Preston, H., Zhang, C., and Marzuki, S. *Biochem. Int.*, **22**, 1067 (1990).
33. Holt, A.J., Harding, A.E., and Morgan-Hughes, J.A. *Nature*, **331**, 717 (1988).
34. Ozawa, T., Tawaka, M., Sugiyama, S., Hattori, K., Ito, T., Ohno, K., Takahashi, A., Sato, W., Takada, G., Mayumi, B., Yamamoto, K., Adachi, K., Koga, Y., and Toshima, H. *Biochem. Biophys. Res. Commun.*, **170**, 830 (1990).
35. Schon, E.A., Rizzuto, R., Moraes, C.T., Nakase, H., Zeviani, M., and DiMauro, S. *Science*, **244**, 346 (1989).
36. Kadenbach, B. and Müller-Höcker, J. *Naturwissenschaften*, **77**, 221 (1990).
37. Moraes, C.T., DiMauro, S., Zeviani, M., Lombes, A., Shanske, S., Miranda, A.F., Nakase, H., Bonilla, E., Werneck, L.C., Servidei, S., Nonaka, I., Koga, Y., Spiro, A.J., Keith, A., Brownell, W., Schmidt, B., Schotland, D.J., Zupang, M., DeVivo, D.C., Schon, E.A., and Rowland, L.P. *N. Engl. J. Med.*, **320**, 1293 (1989).
38. Shoubridge, E.A., Karpati, G., and Hastings, K.E.M. *Cell*, **62**, 43 (1990).
39. Nakase, H., Moraes, C.T., Rizzuto, R., Lombes, A., DiMauro, S., and Schon, E.A. *Am. J. Hum. Genet.*, **46**, 418 (1990).
40. Ikebe, S., Tanaka, M., Ohno, K., Sato, W., Hattori, K., Kondo, T., Mizuno, V., and Ozawa, T. *Biochem. Biophys. Res. Commun.*, **170**, 1044 (1990).
41. Ozawa, T., Tanaka, M., Ikebe, S., Ohno, K., Kondo, T., and Mizuno, Y. *Biochem. Biophys. Res. Commun.*, **172**, 483 (1990).
42. Jarcho, J.A., McKenna, W., Pare, J.A.P., Solomon, S.D., Holocombe, R.F., Dickie, S., Levi, T., Donis-Keller, H., Seidman, J.G., and Seidman, C.E. *N. Engl. J. Med.*, **321**, 1372 (1989).
43. Berko, B.A. and Swift, M. *N. Engl. J. Med.*, **316**, 1386 (1987).
44. Shanske, S., Moraes, C.T., Lombes, A., Miranda, A.F., Bonilla, E., Lewis, P., Whelan, M.A., Ellsworth, C.A., and DiMauro, S. *Neurology*, **40**, 24 (1990).

Mitochondrial DNA Mutations as the Etiology of Cardiomyopathy

TAKAYUKI OZAWA, SATORU SUGIYAMA, AND
MASASHI TANAKA

Department of Biomedical Chemistry, Faculty of Medicine, University of Nagoya, Nagoya 466, Japan

Although cardiomyopathy was considered to be a diagnosis of exclusion, recent advances now permit a definite diagnosis of that condition. Nevertheless, the etiology of cardiomyopathy remains obscure, and its management is still an unresolved problem. Currently, cardiomyopathy is widely accepted as a pluricausal or multifactorial disease. Various factors such as viral infection, free radicals, and altered autonomic nervous function have been proposed to participate in its genesis (*1*). Recently, the role of DNA mutations has been emphasized as the etiology of cardiomyopathy (*2*). The mitochondrial electron transport chain is composed of 4 complexes (complexes I-IV). These complexes together with ATPase are all embedded in the mitochondrial inner membrane and are responsible for the overall process of oxidative phosphorylation, *i.e.*, ATP production as shown in Fig. 1. Mitochondria have their own DNA, 16,569 base pairs (bp) (*3*), which encodes 13 subunits of these 4 complexes (Fig. 1). The rate of mutation of mitochondrial DNA (mtDNA) is expected to be much higher than that of nuclear DNA (*4*), and mtDNA has little intron, suggesting that mutations of mtDNA might be directly linked with deterioration of mitochondrial function. Since mitochondria exclusively produce ATP, mitochondrial dysfunction leads to cellular dysfunction. Using advanced gene technology, we have developed rapid and accurate methods for detection of mutations of mtDNA (*5-7*). In the present paper, using these methods,

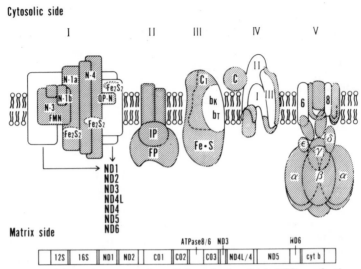

Fig. 1.   Schematic presentation of the mitochondrial ATP production system. Subunits with shadow are encoded by mtDNA, and others by nuclear DNA.

we present evidence that might categorize cardiomyopathy as a mtDNA disease, at least in some cases.

I.   mtDNA ANALYSES

1.   Patients

Cardiac tissue specimens were obtained from seven patients (a 49-year old female, 47-year old male, 53-year old male, 69-year old male, 54-year old female, 40-year old female, and 43-year old male) with hypertrophic or dilated cardiomyopathy of unknown etiology, and from a person who died in an accident as a normal control (a 42-year old female).

2.   Preparation of DNA

The heart muscles (5 mg) were homogenized using a Physcotron Handy Micro Homogenizer (Niti-on, Tokyo, Japan) for 30 sec, and then digested in 1 ml of 10 mM Tris-HCl, 0.1 M EDTA (pH 7.4) containing 0.1 mg/ml proteinase K and 0.5% sodium dodecyl sulfate (SDS). DNA was extracted twice with equal volumes of phenol/chloroform/isoamyl alcohol (25 : 25 : 1), then once with chloroform/isoamyl alcohol (25 : 1). DNA was precipitated with a one-fiftieth volume of 5 M NaCl and two volumes of ethanol at −80°C for 2 hr, and then rinsed with 70% ethanol. The precipitated DNA was recovered in 30 μl of 10 mM Tris-HCl, 0.1 mM EDTA (pH 8.0).

3. Oligonucleotide Primers

Primers used for polymerase chain reaction (PCR) were synthesized using a Shimadzu model NS-1 DNA synthesizer or an Applied Biosystems model 381A DNA synthesizer and then purified on oligonucleotide purification cartridges obtained from Applied Biosystems. The base sequences of the oligonucleotide primers are shown in Table I.

4. Primary PCR Amplification

PCR amplification was carried out on 1 μl of the DNA solution (ca. 10 ng of total DNA) in a final volume of 100 μl which included 200 μM deoxyribonucleoside triphosphate (dNTP), 2.5 units of Taq DNA polymerase (AmpliTaq, Cetus), and PCR buffer (50 mM Tris-HCl, pH 8.4, containing 50 mM KCl, 1.5 mM $MgCl_2$, and 0.01% gelatin) with 1 μM of each primer. For the quantitative analysis of the PCR products, PCR reactions were performed separately using the same template with two different pairs of primers, one of which was used to measure the amount of normal mtDNA, and the other to detect deleted mtDNA. The combi-

TABLE I
Synthesized Primers Used for PCR

Primer[a]	Sequence 5'→3'	Complementary site		
L116	AACTCAAAGGACCTGGCGGT	1,161	to	1,180
L731	TTCATGATTTGAGAAGCCTT	7,311	to	7,330
L853	ACGAAAATCTGTTCGCTTCA	8,531	to	8,550
L881	CACCCAACTATCTATAAACC	8,811	to	8,830
L1167	AACCCCCTGAAGCTTCACCG	11,671	to	11,690
L1641	CGTGAAATCAATATCCCGCA	16,411	to	16,430
H12	GAATCAAAGACAGATACTGC	140	to	121
H38	AAATTTGAAATCTGGTTAGG	400	to	381
H60	AAACATTTTCAGTGTATTGC	620	to	601
H617	CGGGGAAACGCCATATCGGG	6,190	to	6,171
H884	TGCCCGCTCATAAGGGGATG	8,860	to	8,841
H1189	GTTACTAGCACAGAGAGTTC	11,910	to	11,891
H1599	AAATTAGAATCTTAGCTTTG	16,010	to	15,991
H1619	ACTTGCTTGTAAGCATGGG	16,209	to	16,191
H1643	CGAGGAGAGTAGCACTCTTG	16,450	to	16,431

[a]Primers L116, L731, L853, L881, L1167, and L1641 were used for amplification of the light strand of mtDNA. Primers H12, H38, H60, H617, H884, H1189, H1599, H1619, and H1643 were used for amplification of the heavy strand of mtDNA.

TABLE II
Size of Fragments Amplified with Various Combinations of Primers

Combination of primers	Distance between two primers (kb)	Size of amplified fragment (kb)
(Primary PCR)		
L853 + H38	8.4	1.0
L116 + H617	5.0	5.0
(Primer shift PCR)		
L731 + H60	9.9	2.4
L853 + H60	8.7	1.2
L853 + H38	8.4	1.0
(Modified primer shift PCR)		
L853 + H38	8.4	1.0
L853 + H12	8.2	0.7
L853 + H1643	7.9	0.5
L853 + H1619	7.7	0.2
L853 + H1599	7.5	(−)
L853 + H38	8.4	1.0
L881 + H38	8.2	(−)

nations of primers used are shown in Table II. The reactions were carried out for a total of 30 cycles with the use of a Perkin-Elmer/Cetus Thermal Cycler with the following cycle times: denaturation, 15 sec at 94°C; annealing, 15 sec at 50°C; primer extension, 80 sec at 72°C. Amplified fragments were separated by electrophoresis on 1% agarose gels and were detected fluorographically after staining with ethidium bromide.

5. Southern Blot Analysis

Total DNA (100 ng) was digested with 12 units of *Pvu*II and *Pst*I obtained from Toyobo, Osaka, Japan and separated electrophoretically on 0.6% agarose gels. Size standards employed were lambda phage DNA digested with *Hind*III and phage X174 DNA digested with *Hae*III from Nippon Gene, Toyama, Japan. DNA in the gels was denatured and transferred onto Hybond-N+ membrane from Amersham, UK. Hybridization using the PCR-amplified mtDNA fragments as probes was carried out with the Enhanced Chemiluminescence Gene Detection System (ECL kit) from Amersham, UK.

6. Primer Shift PCR (Fig. 2)

In PCR, misannealing of primers sometimes results in amplification of abnormal fragments. In order to ascertain that an amplified fragment is not due to misannealing of primers to an unexpected position of

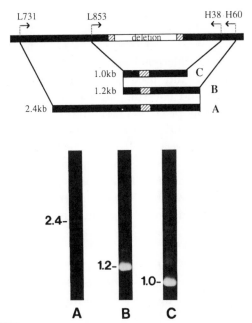

Fig. 2.   The schema and result of the primer shift PCR method. PCR amplification was carried out using primers L731-H60 (lane A), primers L853-H60 (lane B), and primers L853-H38 (lane C). Sizes of amplified fragments are indicated in kb. The shift in the sizes of the amplified fragments parallels the shift in the positions of the primers from L853 to L731 (1.2 kb), and from H60 to H38 (0.2 kb), respectively.

mtDNA, we identified the deletion of mtDNA in the patients by the primer shift method (5), the principle of which is as follows. In the first experiment, a fragment is amplified from the deleted mtDNA using the pair of primers L853 and H60 surrounding the deletion. The size of deletion can be obtained by subtracting the size of the amplified fragment from the distance between the primers. In the second experiment, another fragment is amplified from the deleted mtDNA by another pair of primers L731 and H60. In the third experiment, the third fragment is amplified from the deleted mtDNA using the third pair of primers L853 and H38. The shift in the sizes of the amplified fragments should parallel the shift in the positions of the primers from L853 to L731 (1.2 kb), and from H60 to H38 (0.2 kb), respectively. If the deletion sizes calculated from the two experiments are identical, it is concluded that the amplified fragments are not due to misannealing of the primers but due to the presence of the deleted mtDNA.

7. PCR-Southern Analysis (Fig. 3)

PCR-Southern analysis is another method (6) to confirm the existence of mtDNA deletion. Three different fragments were amplified by PCR from the normal mtDNA and used as the probes: probe A, spanning positions 8,531–8,860 and covering the left-side boundary of the deletion, was amplified using primers L853 and H884; probe B, spanning positions 11,671–11,910 and covering the middle part of the deletion, was amplified using primers L1167 and H1189; probe C, spanning positions 16,411–140 and covering the right-side boundary of the deletion, was amplified using primers L1641 and H12. When the abnormal PCR fragment was derived from the deleted mtDNA, probes A and C would hybridize to the abnormal fragment, but probe B would not hybridize to the fragment.

8. Modified Primer Shift PCR Method (Fig. 4)

To estimate the deleted region efficiently and precisely, we modified the primer shift method. If the complementary site of the primer used for

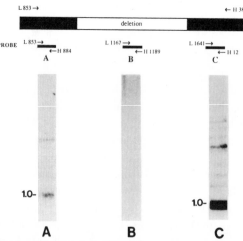

Fig. 3. The schema and result of the PCR-Southern method. Three different fragments were amplified by PCR from the normal mtDNA and used as the probes: probe A (spanning positions 8,531-8,860), amplified with primers L853 and H884: probe B (spanning positions 11,671-11,910), amplified with primers L1167 and H1189; and probe C (spanning positions 16,411-140), amplified with primers L1641 and H12. Probes A and C hybridized to small fragments, but probe B did not hybridize to those fragments.

PCR is outside the deletion, a fragment can be produced. When the complementary site of the primer is inside the deletion, no fragment can be amplified. Therefore, the deletion is estimated to start between the site of the primer which produces a fragment and that of the primer which produces no fragment.

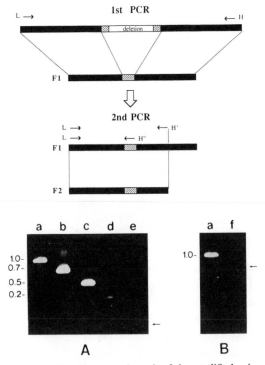

Fig. 4.   The schema and result of the modified primer shift PCR method. In the first experiment, fragment F1 is amplified from the deleted mtDNA with primers L and H. In the second experiment, PCR is performed with primers L and H′ using fragment F1 as the template. If the complementary site of primer H′ is outside the deletion, fragment F2 can be produced. If the complementary site of primer H″ is inside the deletion, no fragment can be amplified. Therefore, the deletion is estimated to start between the sites of primers H′ and H″. As shown in the upper part of the figure, PCR reamplification was carried out using primers L853-H38 (lane A-a, lane B-a), primers L853-H12 (lane A-b), primers L853-H1643 (lane A-c), primers L853-H1619 (lane A-d), primers L853-H1599 (lane A-e), and primers L881-H38 (lane B-f). Sizes of amplified fragments are indicated in kb. A fragment is eliminated by shifting the primer from H1619 (lane A-d) to H1599 (lane A-e). Similarly, a fragment is eliminated by shifting the primer from L853 (lane B-a) to L881 (lane B-f), as shown in the lower part of the figure.

9. Asymmetric PCR Amplification

PCR reamplification was carried out on 2.5 μl of the primary PCR product in a final volume of 100 μl which included the reagents described above with 0.01 μM of one primer and 1 μM of another primer essentially according to the method of Gyllensten and Erlich (8). The combination of primers used for patients is shown in Table III. PCR was performed for a total of 35 cycles as above. The PCR product containing single-stranded DNA was precipitated with a 0.6 volume of 20% polyethylene-glycol 6,000 (PEG) containing 2.5 M NaCl essentially according to the method of Perbal (9). After incubation at 4°C for 1 hr, the mixture was centrifuged at 11,000 $\times g$ for 10 min. The precipitate was suspended in 100 μl of 7.5% PEG containing 0.94 M NaCl and centrifuged as above. The pellet was rinsed with 0.5 ml of 70% ethanol, dried in vacuum for 15 min, and dissolved in 10 μl of distilled water. The removal of dNTP and primers by this PEG precipitation step is essential for reduction of the background in the subsequent sequencing.

10. DNA Sequencing

DNA was sequenced by Sanger's dideoxynucleotide chain termination method using the incorporation of α-[^{32}P]dCTP as the radiolabeling extension method (7). Sequenase reactions were performed using a kit supplied by United States Biochemicals. For the labeling reaction, the amplified single-stranded DNA-enriched template (7.5 μl, 0.5–1.0 pmol) was mixed with 1 μl of 10 μM sequencing primer listed in Table III, and the primer-template mixture was heated to 100°C for 10 min and immediately placed on ice. Reactions were initiated by adding 8.8 μl of this mixture to 5.2 μl of dideoxy G, A, T, and C reaction mixtures composed of reagents provided in the "Sequence" kit as follows: 2 μl of five-times diluted labeling mix, 0.25 μl of Sequenase (3.25 units), 1.75 μl of "dilution buffer," 2.2 μl of "5 $\times$ Sequenase buffer," and 1 μl of 0.1M DTT. The mixture was incubated at 37°C for 2 min. The product (3.5 μl) was transferred to four tubes containing 2.5 μl of one of four "termination

TABLE III
Combination of Primers Used for Asymmetric Amplification of mtDNA and for Sequencing of the Templates

Primers for amplification	Primer for sequencing
L853 (0.01)+H38 (1)	L853

Figures in parentheses are concentration of primers in μM used for asymmetric amplification.

mixes" and incubated at 37°C for 2 min. After addition of 4 μl of "stop solution," the mixture was heated to 100°C for 3 min and 3 μl was loaded onto a 6% polyacrylamide/7 M urea sequencing gel.

II. mtDNA MUTATIONS IN PATIENTS WITH CARDIOMYOPATHY

Figure 5 shows the PCR amplification of mtDNA from the heart muscle using primers L853 and H38. Multiple abnormal fragments which were derived from a deleted mtDNA were detected in patients with cardiomyopathy, and the 1.0 kbp fragment was commonly observed. Coincident with the distance 1.2 kbp between L853 and L731, a fragment of 1.2 kbp in lane B was shifted to 2.4 kbp in lane A (Fig. 2). Using the other primer pair, L853-H38, 0.2 kbp shorter than the pair used in lane B, the 1.2 kbp fragment was shifted to 1.0 kbp as shown in lane C (Fig. 2). These results clearly indicate that the PCR amplified abnormal fragments were not the products of misannealing of PCR primer(s) to mtDNA, but the fragments derived from real deletions of mtDNA. As shown in Fig. 3, both probes A and C hybridized to several abnormal fragments; probe B did not hybridize to those fragments. These results demonstrate that the probe B region is really deleted in these mtDNAs in

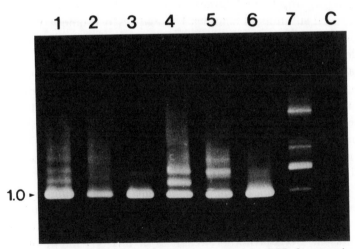

Fig. 5. Mitochondrial DNA fragments amplified by PCR from specimens obtained from patients with cardiomyopathy. In mtDNA fragments obtained from patients 1–7 (lanes 1–7), multiple abnormal fragments were detected, whereas no fragments existed in the control specimen (lane C).

patients with cardiomyopathy. Before sequencing, we determined precisely the location of the deletion using the modified primer shift method. As shown in Fig. 4, a fragment was eliminated by shifting the primer from H1619 (Fig. 4, A-d) to H1599 (Fig. 4, A-e). Similarly, a fragment was eliminated by shifting the primer from L853 (Fig. 4, B-a) to L881 (Fig. 4, B-f). We could deduce from these results that the right-side boundary of this deletion was located between position 15,991 and position 16,210 of normal mtDNA, and the left-side boundary was located between position 8,531 and position 8,830. We therefore sequenced this region. The direct repeats of this fragment are presented in Fig. 6. The crossover sequence was demonstrated to be a 12 bp directly repeated sequence of 5'-CATC-AACAACCG-3', which was located on the boundaries of the deletion between the ATPase 6 gene and the D-loop region. The deletion spanned 7,436 bp.

We examined total mtDNA sequences in another patient with hypertrophic cardiomyopathy and, as shown in Table IV, observed multiple point mutations.

III.   ROLE OF mtDNA MUTATIONS IN CARDIOMYOPATHY

The human mitochondrial genome contains protein-coding genes specifying hydrophobic subunits of the mitochondrial electron transport chain: seven subunits of complex I, the apocytochrome b of complex III, and three subunits of complex IV. Two subunits of complex V are also encoded in mtDNA. Together with other protein subunits encoded by nuclear DNA and synthesized in and imported from the extramitochondrial cytoplasm, these mitochondrial translation products are assembled into functional enzyme complexes of the respiratory chain. The rest of the mitochondrial genome contains genetic information essential to the assembly of the mitochondrial protein-synthesizing machinery required for the expression of the protein-coding genes: genes specifying two mitochondrial rRNAs and 22 organelle specific tRNAs.

Because mtDNA has neither histone nor a repair system and is constantly exposed to oxygen radicals leaked from the mitochondrial electron-transport chain, it is likely to cause mutations at a higher rate than nuclear DNA. Another weakness of mtDNA might be due to its highly economical package; the expression of the whole mitochondrial genome is needed for maintenance of the mitochondrial energy transducing system, whereas only about 7% of the nuclear genome is ever

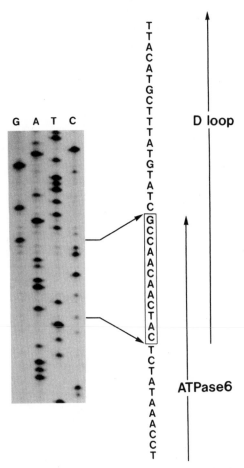

Fig. 6.   Direct sequencing of the deleted mitochondrial DNA from heart muscle. Shown is a portion of an autoradiograph of sequencing gel of the amplified DNA from the deleted mitochondrial DNA. A mitochondrial DNA fragment with 7,436 bp deletion was sequenced and the direct repeat was identified as 5'-CATCAA-CAACCG-3', which was located in both the ATPase 6 gene and the D-loop region.

expressed at any particular differentiated stage (*10*). Thus, while it is probable that a mutational event in the nuclear DNA will affect a non-expressed region of the genome, any mutation in the mtDNA will involve a functionally important part of the genome.

Keeping in mind the importance of mitochondria in the cellular energy metabolism, mitochondrial dysfunction due to mtDNA mutations

TABLE IV
Point Mutations in mtDNA of a Patient with Hypertrophic Cardiomyopathy (HCM)

No.	Region		Std.	Mut.	Std.	Mut.	N/C
tRNA							
3243	tRNA-Leu		A	G			
4386	tRNA-Gln		T	C			
8308	tRNA-Lys		A	G			
Amino acid	Change						
8906	ATP6	2	A	G	His	Arg	C
10398	ND3	1	A	G	Thr	Ala	C
11084	ND4	1	A	G	Thr	Ala	C
3338	ND1	2	T	C	Val	Ala	N
8701	ATP6	1	A	G	Thr	Ala	N
14199	ND6	1	G	T	Pro	Thr	N
14272	ND6	3	G	C	Phe	Leu	N
14368	ND6	3	G	C	Phe	Leu	N
15326	Cytb	1	A	G	Thr	Ala	N
Amino acid	No change						
3423	ND1	3	G	T	Val		
3657	ND1	3	C	A	Leu		
4769	ND2	3	A	G	Met		
4958	ND2	3	A	G	Met		
4985	ND2	3	G	A	Gln		
6455	CO1	3	C	T	Phe		
7028	CO1	3	C	T	Ala		
8200	CO2	3	T	C	Ser		
9251	CO3	3	A	G	Pro		
9540	CO3	1	T	C	Leu		
9824	CO3	3	T	C	Leu		
10400	ND3	3	C	T	Thr		
10873	ND4	3	T	C	Pro		
11017	ND4	3	T	C	Ser		
11335	ND4	3	T	C	Asn		
11719	ND4	3	G	A	Gly		
12705	ND5	3	C	T	Ile		
12771	ND5	3	G	A	Glu		
14364	ND6	1	G	A	Leu		
14365	ND6	3	G	C	Val		
14783	Cytb	1	T	C	Leu		
15043	Cytb	3	G	A	Gly		
15301	Cytb	3	G	A	Leu		

might be a major cause of the modifications that occur in cardiomyopathy. The sequence of the deleted mtDNA produces a stop codon 77 nucleotides distant from the boundary, resulting in premature termination of the D-loop region. Therefore, the sequence of mutant mtDNA predicts a 7.5-kDa abnormal protein composed of 41 amino acid residues

No.	Region	Std.	Mut.	Std.	Mut.	N/C
D-Loop						
73	D-Loop	A	G			
106	D-Loop	G	Del.			
107	D-Loop	G	Del.			
108	D-Loop	A	Del.			
109	D-Loop	G	Del.			
110	D-Loop	C	Del.			
111	D-Loop	A	Del.			
16209	D-Loop	T	C			
16223	D-Loop	C	T			
16324	D-Loop	T	C			
rRNA						
750	12SrRNA	A	G			
1438	12SrRNA	A	G			
2626	16SrRNA	T	C			
2706	16SrRNA	A	G			
2772	16SrRNA	C	T			
3106	16SrRNA	C	Del.			
NCR						
200	NCR	A	G			
263	NCR	A	G			
303.1	NCR	—	C			
311.1	NCR	—	C			
489	NCR	T	C			
514	NCR	C	Del.			
515	NCR	A	Del.			
5582	NCR	A	G			
5894.1	NCR	—	C			

Std.: standard mtDNA sequences reported by Anderson *et al.* (*3*). Mut.: mutation. C: conserved among human, bovine, and rodent. N: non-conserved. Del.: deleted. NCR: non-coding region.

from the N-terminal side of the ATPase subunit 6 and of 25 amino acid residues from the sequence of the D-loop region. It is possible that this mutant protein might disturb the molecular assembly of the energy-transducing complexes, though further study must be awaited to clarify whether or not the mutant mRNA can be translated in the mitochondria. MtDNA mutations are revealed to be responsible for the development of pathological conditions such as Kerns-Sayre syndrome (*11*), chronic progressive external ophthalmoplegia (*11*, *12*), myoclonus epilepsy with ragged-red fibers (*13*), Leber's hereditary optic neuropathy (*14*, *15*), mitochondrial myopathy, encephalopathy, lactic acidosis, and stroke-like episodes (*16*, *17*), and Parkinson's disease (*18*, *19*). These diseases should now be categorized as "mtDNA diseases". I would like to propose here that cardiomyopathy might, at least in part, be categorized as a "mtDNA disease".

To mitigate the inhibition of the energy metabolism, we proposed redox therapy (20). Mitochondria produce ATP using energy transduced from exogenic redox reactions. Bypassing blockage in the mitochondrial energy producing system by using substances which have redox potentials permitting interaction with relevant complexes in this system might mitigate disturbances of mitochondrial energy production caused by mtDNA mutations. Although the rate of electron flow by redox therapy now available is much lower than that of the normal activity of the mitochondrial electron transport system, the beneficial effects of redox therapy have been confirmed (21). To improve the efficacy of this kind of therapy, the development of substances which have even more suitable redox properties is anticipated.

Finally, we would like to point out that both familial and sporadic cardiomyopathy associated with mtDNA mutations might occur because mtDNA and mtDNA deletions are both inherited maternally (12), and it is possible that non-inherited, i.e., acquired deletion of mtDNA might also occur.

SUMMARY

Mitochondria, bio-engines, exclusively produce ATP. Mitochondria contain an electron transport chain which is composed of 4 complexes, and each complex consists of various numbers of subunits. These complexes together with ATPase are responsible for ATP production. Mitochondria contain their own DNA (mtDNA), and some subunits are encoded by mtDNA. Accumulating evidence emphasizes the role of genetic factors in the pathogenesis of cardiomyopathy. The rate of mtDNA mutations is estimated to be much higher than that of nuclear DNA, and mutations of mtDNA are involved in the basis of various diseases. Recent advances in gene technology, especially in PCR make it possible to analyze mtDNA mutations in a small quantity of tissue. We developed various rapid and accurate methods to detect these mutations and demonstrated that multiple mitochondrial mutations exist in patients with cardiomyopathy. One mutation was based on the facts that the directly repeated 5'-CATCAACAACCG-3' sequence is found in both the ATPase6 gene and the D-loop region, and that pseudo-recombination between directly repeated sequences results in a 7.4 kbp deletion. We also demonstrated that multiple point mutations exist in patients with hypertrophic cardiomyopathy. Some subunits of the mitochondrial ATP production system could not be biosynthesized by mtDNA with mutations. As a result, energy production might be inhibited, which could be linked to the genesis of cardiomyopathy.

Acknowledgment

We express our appreciation to Michael Bodman, a language consultant of our department, for reading the previous draft and making suggestions on language and style.

REFERENCES

1. Wynne, J. and Braunwald, E. *In* "Heart Disease," Vol. II, ed. E. Braunwald, p. 1410 (1988). W.B. Saunders, Philadelphia.
2. Burn, J. *In* "Progress in Cardiology 2/2," ed. D.P. Zipes and D.J. Rowlands, p. 45 (1989). Lea & Febiger, Philadelphia.
3. Anderson, S., Bankier, A.T., Barrell, B.G., de Bruijn, M.H.L., Coulson, A.R., Drouin, J., Eperon, I.C., Nierlich, D.P., Roe, B.A., Sanger, F., Schreier, P.H., Smith, A.J.H., Staden, R., and Young, I.G. *Nature*, **290**, 457 (1981).
4. Linnane, A.W., Marzuki, S., Ozawa, T., and Tanaka, M. *Lancet*, **i**, 642 (1989).
5. Sato, W., Tanaka, M., Ohno, K., Yamamoto, T., Takada, G., and Ozawa, T. *Biochem. Biophys. Res. Commun.*, **162**, 664 (1989).
6. Ozawa, T., Tanaka, M., Sugiyama, S., Hattori, K., Ito, T., Ohno, K., Takahashi, A., Sato, W., Takada, G., Mayumi, B., Yamamoto, K., Adachi, K., Koga, Y., and Toshima, H. *Biochem. Biophys. Res. Commun.*, **170**, 830 (1990).
7. Tanaka, M., Sato, W., Ohno, K., Yamamoto, T., and Ozawa, T. *Biochem. Biophys. Res. Commun.*, **164**, 156 (1989).
8. Gyllensten, U.B. and Erlich, H.A. *Proc. Natl. Acad. Sci. U.S.A.*, **85**, 7652 (1988).
9. Perbal, B. *In* "A Practical Guide to Molecular Cloning," ed. B. Perbal, p. 636 (1988). John Wiley & Sons, New York.
10. Watson, J.D., Hopkins, N.H., Roberts, J.W., Steitz, J.A., and Weiner, A.M. *In* "Molecular Biology of the Gene," 4th ed. Vol. 1, p. 691 (1987). Benjamin Cummings, Menlo Park, California.
11. Moraes, C.T., DiMauro, S., Zeviani, M., Lombes, A., Shanske, S., Miranda, A.F., Nakase, H., Bonilla, E., Werneck, L.C., Servidei, S., Nonaka, I., Koga, Y., Spiro, A.J., Brownell, A.K.W., Schmidt, B., Schotland, D.L., Zupanc, M., DeVivo, D.C., Schon, E.A., and Rowland, L.P. *N. Engl. J. Med.*, **320**, 1293 (1989).
12. Ozawa, T., Yoneda, M., Tanaka, M., Ohno, K., Sato, W., Suzuki, H., Nishikimi, M., Yamamoto, M., Nonaka, I., and Horai, S. *Biochem. Biophys. Res. Commun.*, **154**, 1240 (1988).
13. Yoneda, M., Tanno, Y., Horai, S., Ozawa, T., Miyatake, T., and Tsuji, S. *Biochem. Int.*, **21**, 789 (1990).
14. Wallace, D.C., Singh, G., Lott, M.T., Hodge, J.A., Shurr, T.G., and Lezza, A.M.S.,

Elsas, II L.J., and Nikoskelainen, E.K. *Science*, **242**, 1427 (1988).

15. Yoneda, M., Tsuji, S., Yamauchi, T., Inuzuka, T., Miyatake, T., Horai, S., and Ozawa, T. *Lancet*, **i**, 1076 (1989).
16. Ino, H., Tanaka, M., Ohno, K., Hattori, K., Ikebe, S., Sano, T., Ozawa, T., Ichiki, T., Kobayashi, M., and Wada, Y. *Lancet*, **i**, 234 (1991).
17. Tanaka, M., Ino, H., Ohno, K., Ohbayashi, T., Ikebe, S., Sano, T., Ichiki, T., Kobayashi, M., Wada, Y., and Ozawa, T. *Biochem. Biophys. Res. Commun.*, **174**, 861 (1991).
18. Ikebe, S., Tanaka, M., Ohno, K., Sato, W., Hattori, K., Kondo, T., Mizuno, Y., and Ozawa, T. *Biochem. Biophys. Res. Commun.*, **170**, 1044 (1990).
19. Ozawa, T., Tanaka, M., Ikebe, S., Ohno, K., Kondo, T., and Mizuno, Y. *Biochem. Biophys. Res. Commun.*, **172**, 483 (1990).
20. Ozawa, T., Tanaka, M., Suzuki, H., and Nishikimi, M. *Brain Dev.*, **9**, 76 (1987).
21. Jinnai, K., Yamada, H., Kanda, F., Masui, Y., Tanaka, M., Ozawa, T., and Fujita, T. *Eur. Neurol.*, **30**, 56 (1990).

Kinetic Mechanism for Activation of Sarcoplasmic Reticulum Ca^{2+}-ATPase by Ca^{2+} and ATP

MUNEKAZU SHIGEKAWA, SHIGEO WAKABAYASHI, AND TAROU OGURUSU

Department of Molecular Physiology, National Cardiovascular Center Research Institute, Suita, Osaka 565, Japan

Ca^{2+} transport across the sarcoplasmic reticulum (SR) is catalyzed by an integral membrane protein, Ca^{2+}-ATPase (*1, 2*), which consists of a single polypeptide with M.W.$=110$ kDa. It has a highly asymmetric structure with respect to the membrane in that a major portion ($\sim 60\%$) of its mass protrudes into the cytoplasm as a pear-shaped head with a narrow stalk, while the rest is buried within the membrane (*3–5*). Recent chemical labeling, site-directed mutagenesis and fluorescence energy transfer studies (*4–7*) have provided evidence that the ATP-binding site is located in the cytoplasmic portion of the enzyme, whereas the Ca^{2+} binding site is formed from its transmembrane segments. Because these two functional sites are presumably separated from each other, coupling of ATP hydrolysis and Ca^{2+} movement probably occurs indirectly through the structural change of the ATPase molecule. At present, however, precise information is not available concerning the physical locations of these sites in the three dimensional structure of the enzyme and their interactions during Ca^{2+} transport. Therefore, the physical mechanism of Ca^{2+} transport is not known.

In this brief review, we would like to discuss the kinetic aspects of the mechanism for Ca^{2+} transport by the Ca^{2+}-ATPase, since its structural aspect will be treated elsewhere in this book. In particular, we will concentrate on the mechanism for activation of Ca^{2+}-ATPase by Ca^{2+} and ATP, which is an important initial step in the Ca^{2+} transport.

I. REACTION SEQUENCE

Figure 1 shows a simplified version of reaction cycle for Ca^{2+}-ATPase (*1, 2, 8*). Binding of two Ca^{2+} ions to the high-affinity Ca^{2+} site from the cytoplasmic side of the SR membrane induces phosphorylation of the enzyme by ATP to form the ADP-sensitive phosphoenzyme (E_1P). Ca^{2+} ions in E_1P are occluded in that they cannot be exchanged with Ca^{2+} ions in the aqueous medium on either side of the SR membrane. E_1P is then converted spontaneously to the ADP-insensitive phosphoenzyme (E_2P) and, at the same time, Ca^{2+} ions are exposed on the luminal surface and released. E_2P is subsequently hydrolyzed to yield a dephosphoenzyme, which then starts another round of the reaction cycle. The ATPase reaction can be reversed, E_2P being formed from inorganic phosphate, when the Ca^{2+} concentration on the cytoplasmic surface is very low. Vanadate reacts with the Ca^{2+}-free dephosphoenzyme as an analogue for inorganic phosphate.

The Ca^{2+}-ATPase utilizes a magnesium-ATP complex as a physiological substrate (*9*). It is well known that Mg^{2+} is required for the rapid turnover of Ca^{2+}-ATPase. Recent divalent cation-binding studies (*10, 11*) provided direct evidence that Mg^{2+} derived from the magnesium-ATP complex remains bound to the Ca^{2+}-ATPase (probably at the catalytic center) at least until E_2P is hydrolyzed (Fig. 1), and that binding of this magnesium is responsible for induction of the high rate of enzyme turnover.

The molecular mechanism by which ATP hydrolysis and translocation of two Ca^{2+} ions are coupled in the Ca^{2+}-ATPase is not known. It is widely believed that the free energy of ATP is utilized to induce a conformational change in the phosphoenzyme, which causes vectorial

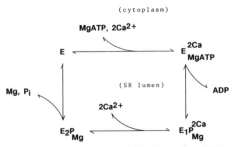

Fig. 1.   Reaction cycle of SR Ca pump ATPase.

movement of the transported Ca^{2+} ions as well as a change in affinity of the Ca^{2+} site for transported Ca^{2+} ions. So, some form of a gating mechanism is activated at the Ca^{2+} site in the membrane by the structural perturbation generated by phosphorylation at the ATP processing site in the cytosolic portion of the enzyme. It was proposed that the Ca^{2+}-ATPase exists in two major conformations which alternate during the reaction cycle (1). This two-state model assumes that the Ca^{2+} site in one state faces the cytoplasm and exhibits high affinity for Ca^{2+}, while the Ca^{2+} site in the other state faces the lumen and exhibits low affinity for Ca^{2+}. Conversion between these two states of the Ca^{2+} site in the phosphorylated and unphosphorylated enzymes may permit simultaneous changes in the Ca^{2+} affinity and the sidedness of the Ca^{2+} site to occur cyclically. Although this two-state model has commonly been used to explain experimental results, there has so far been no firm experimental evidence showing that the Ca^{2+} site changes its sidedness and its Ca^{2+} affinity simultaneously.

II. BINDING OF Ca²⁺ TO Ca²⁺-ATPase

Interaction of Ca^{2+} with unphosphorylated Ca^{2+}-ATPase has been analyzed in equilibrium as well as in transient kinetic measurements. Equilibrium measurements revealed that the stoichiometry for high-affinity Ca^{2+} binding is 2 mol per mol of the phosphorylation site and that binding exhibits positive cooperativity with a $K_{1/2}$ value of 0.5–5 μM (12– 14, 34) (see inset of Fig. 2).

Binding of Ca^{2+} to the Ca^{2+} site induces a change in the enzyme conformation, which has been detected by changes in various spectral properties, tryptic digestion pattern, and chemical reactivity of amino acid residues of the enzyme (14– 20). These Ca^{2+}-induced conformational changes involve neither significant changes in the content of secondary structures of the enzyme (21, 22) nor a change in the intramolecular distance determined by energy transfer between fluorescein 5′-isothiocyanate (FITC) and 5-[[2-[(iodoacetyl)amino]ethyl]amino]naphthalene-1-sulfonic acid (IAEDANS) labels incorporated into Lys-515 and probably Cys-674, respectively, in the cytoplasmic portion of the enzyme molecule (23). Thus the Ca^{2+} binding induces a change in the tertiary structure of the ATPase but may not significantly affect the overall shape of the ATPase molecule.

Transient kinetics of binding and dissociation of Ca^{2+} have been

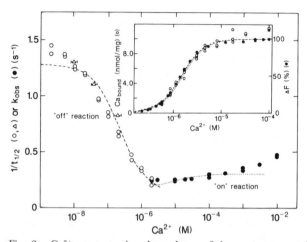

Fig. 2. Ca^{2+} concentration dependence of the rate constants for the on and off fluorescence reactions. The rate constants for the NBD fluorescence rise following addition of Ca^{2+} to the Ca^{2+}-free Ca^{2+}-ATPase (●) and for the decrease in NBD fluorescence following addition of EGTA (○) or quin 2 (△) to the Ca^{2+}-bound enzyme are plotted as a function of the final concentration of ionized Ca^{2+}. The off rate constant is described as the reciprocal of the time at which half the maximal decrease in fluorescence occurs. The inset shows the Ca^{2+} dependence of the equilibrium levels of Ca^{2+} binding to the Ca^{2+}-ATPase and NBD fluorescence. Experimental conditions: 0.1 M KCl, 2 mM $MgCl_2$, pH 6.5, and 11°C (for details, see ref. 32).

studied by following the change in intrinsic protein fluorescence (24–27), the Ca^{2+}-dependent enzyme phosphorylation by ATP (28, 29), and direct binding of radioactive calcium to the enzyme (25, 30). These data provided evidence that two binding sites for Ca^{2+} ions apparently interact and that Ca^{2+} ions bind to these sites in a strictly sequential manner (cf. Fig. 3). These previous data, however, have not clarified the precise relationship between Ca^{2+} binding and the enzyme conformational change. This is because only the intrinsic protein fluorescence has been available as a conformational probe with which to study the rapid conformational transitions associated with binding and dissociation of Ca^{2+}.

We recently introduced a fluorescent 4-nitrobenzo-2-oxa-1,3-diazole (NBD) label into a specific sulfhydryl residue (Cys-344) near the phosphorylation site (Asp-351) of the Ca^{2+}-ATPase (31). The NBD label is thus probably located in the cytoplasmic portion of the enzyme (4). Ca^{2+} binding to the enzyme induces up to a two-fold increase in the intensity of the NBD fluorescence, and Ca^{2+} binding and the NBD fluorescence

Scheme I

$$E_2 \; \overset{Ca^{2+}}{\longleftrightarrow} \; E_2Ca \; \overset{Ca^{2+}}{\longleftrightarrow} \; E_2Ca_2 \; \longleftrightarrow \; E_1Ca_2$$

Scheme II

$$E_2 \; \underset{k_{-1}}{\overset{k_1\,[Ca^{2+}]}{\rightleftharpoons}} \; E_2Ca \; \underset{k_{-2}}{\overset{k_2}{\rightleftharpoons}} \; E_1Ca \; \underset{k_{-3}}{\overset{k_3[Ca^{2+}]}{\rightleftharpoons}} \; E_1Ca_2$$

Scheme III

$$E_2 \; \underset{k_{-1}}{\overset{k_1}{\rightleftharpoons}} \; E_1 \; \underset{k_{-2}}{\overset{k_2\,[Ca^{2+}]}{\rightleftharpoons}} \; E_1Ca \; \underset{k_{-3}}{\overset{k_3[Ca^{2+}]}{\rightleftharpoons}} \; E_1Ca_2$$

Fig. 3.   Mechanisms for Ca²⁺ binding to Ca²⁺-ATPase.

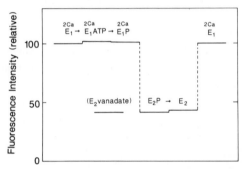

Fig. 4.   A schematic diagram of the relative fluorescence levels of reaction intermediates of Ca²⁺-ATPase at pH 6.0. For experimental details, see ref. *31.*

exhibit the same Ca²⁺ concentration dependence under equilibrium
conditions (Fig. 2, inset). Furthermore, at low pH, the enzyme states can
be grouped into two categories, one being in a high-fluorescence state (the
unphosphorylated enzyme with bound calcium, the enzyme-ATP-calci-
um complex, and the ADP-sensitive phosphoenzyme (E_1P)) and the other
being in a low-fluorescence state (the unphosphorylated enzyme without
bound calcium, the ADP-insensitive phosphoenzyme (E_2P), and the
enzyme-vanadate complex)) (Fig. 4). This behavior of NBD fluorescence
is seemingly compatible with the hypothesis that two major enzyme
conformations alternate during the reaction cycle (see above).

We have used NBD fluorescence as a probe to monitor the conformational transition in the Ca²⁺-ATPase and directly compared the time
courses of the NBD fluorescence change with those of binding and release
of the Ca²⁺ ions from the Ca²⁺ site on the NBD-labeled unphosphorylat-

ed enzyme (32). The following are the results obtained under the conditions of pH 6.5, 2 mM Mg^{2+} and 11°C: (a) the NBD fluorescence rise upon addition of Ca^{2+} (1–100 μM) is exponential, and its k_{obs} value is almost independent of added Ca^{2+} concentration (Fig. 2); (b) the rates of Ca^{2+} binding to the NBD-enzyme at Ca^{2+} concentrations from 2 to 100 μM are slow and almost equal to those for the NBD fluorescence rise measured under the same conditions; (c) neither the NBD fluorescence decrease nor dissociation of Ca^{2+} from the Ca^{2+}-bound enzyme proceeds monoexponentially except when the free Ca^{2+} concentration in the reaction medium is very low ($< \sim 10$ nM). Under the latter conditions, the NBD fluorescence decrease, but not the Ca^{2+} dissociation, is preceded by an initial induction period; (d) Ca^{2+} dissociation from the Ca^{2+}-bound enzyme proceeds at a significantly faster rate than the NBD fluorescence decrease.

The results summarized in the preceding paragraph can be explained by using one of three simple reaction mechanisms shown in Fig. 3, in which binding and dissociation of Ca^{2+} ions are correlated in different ways with the conformational transition between two enzyme states having low and high levels of NBD-fluorescence. The results are compatible only with the mechanism in scheme III of Fig. 3 and a set of kinetic parameters listed in Table I. Scheme III shows that two Ca^{2+} ions bind to the unphosphorylated Ca^{2+}-ATPase sequentially only after the conformational change detectable by the NBD fluorescence. Table I shows that the first and second Ca^{2+} ions bind with high affinity (K_d, 0.1–0.4 μM) to the high-fluorescence state of the enzyme (E_1) and that the equilibrium between the high-fluorescence (E_1) and low-fluorescence (E_2) states in the absence of Ca^{2+} is shifted mostly to the E_2 state. In contrast, the E_2 state does not bind Ca^{2+} with high affinity.

Intrinsic protein fluorescence has long been considered to be a good marker of the Ca^{2+}-induced conformational transition in the unphosphor-

TABLE I
Parameters Describing Ca^{2+} Binding to Ca^{2+}-ATPase

Parameter	Value
k_1	0.3 sec^{-1}
k_{-1}	20 sec^{-1}
k_2	8×10^6 M^{-1} sec^{-1}
k_{-2}	1.6 sec^{-1}
k_3	3.5×10^7 M^{-1} sec^{-1}
k_{-3}	5 sec^{-1}

ylated Ca^{2+}-ATPase. The Ca^{2+} concentration dependence of a rise of intrinsic fluorescence is identical to that of Ca^{2+} binding to the enzyme (*25*). Recent evidence has shown that intrinsic fluorescence monitors fluorescence of tryptophan residues located mostly on the transmembrane region of the Ca^{2+}-ATPase (*33*).

As might be inferred from different locations of the fluorescence probes, intrinsic fluorescence and NBD fluorescence seem to monitor different conformational changes of Ca^{2+}-ATPase. Ca^{2+} binding at neutral pH in the presence of millimolar Mg^{2+} produces a biphasic signal for the intrinsic fluorescence (*26*), whereas it produces a monoexponential signal for the NBD fluorescence (*32*). In addition, the Ca^{2+}-induced NBD fluorescence rise in the unphosphorylated enzyme at pH 6.0 can be reversed completely when Ca^{2+} ions are released from the phosphorylated enzyme (formation of E_2P, see Figs. 1 and 4). In contrast, corresponding reversal of the Ca^{2+}-induced rise of the intrinsic fluorescence does not occur upon release of Ca^{2+} ions from the phosphorylated enzyme (*27*, *33*).

It is possible that the Ca^{2+}-induced intrinsic fluorescence signal primarily monitors the conformational change in the transmembrane domain surrounding the Ca^{2+} site. The NBD-fluorescence, on the other hand, cannot monitor the local event at the Ca^{2+} site because the kinetics of Ca^{2+} dissociation and NBD fluorescence decrease are clearly different (see above), and because the NBD label is located in the cytoplasmic, phosphorylation domain of the Ca^{2+}-ATPase (Cys-344). The NBD signal thus appears to monitor a long-range conformational change that reflects the interaction between the different functional domains of Ca^{2+}-ATPase.

III.   ACTIVATION OF Ca^{2+}-ATPase BY Ca^{2+} AND ATP

The rate of rise of the intrinsic tryptophan fluorescence following addition of Ca^{2+} to the Ca^{2+}-ATPase is much slower than the rate of phosphorylation of the enzyme induced by the addition of both Ca^{2+} and ATP (*14*). This suggests that ATP accelerates enzyme activation by Ca^{2+}. This accelerating effect is evident in Fig. 5, in which the rates of Ca^{2+} binding to the unphosphorylated NBD-enzyme and the accompanying NBD fluorescence rise are measured.

In the absence of Ca^{2+}, the Ca^{2+}-ATPase binds MgATP with high affinity (K_d, 3–5 μM) at a stoichiometric ratio of 1 mol per mol of the phosphorylation site (*13*, *34*). This ATP binding induces fast increases of

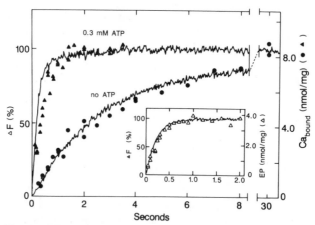

Fig. 5. Effect of ATP on time courses of Ca^{2+} binding and the rise in NBD fluorescence. The inset shows time courses for the rise in NBD fluorescence and the formation of phosphoenzyme. Conditions: 0.1 M KCl, 2 mM $MgCl_2$, 50 μM $CaCl_2$, 0 or 0.3 mM ATP, pH 6.5 and 11°C (for details, see ref. 32).

both intrinsic protein fluorescence and NBD fluorescence (27, 31, 35). The kinetic analysis of the ATP-induced NBD fluorescence changes in the presence of various concentrations of Ca^{2+} revealed that scheme III of Fig. 3 can also be used to describe the mechanism for Ca^{2+} binding in the presence of ATP, and that the ATP-induced fast NBD fluorescence rise detected in the Ca^{2+}-free enzyme monitors the formation of the high-fluorescence state, E_1, which is also an obligatory intermediate for Ca^{2+} binding in the presence of ATP (see Fig. 6 for the mechanism of Ca^{2+} binding in the presence of ATP).

Fitting of the NBD fluorescence data to the mechanism of Fig. 6 gave values for the kinetic parameters listed in the same figure (32). According to these values, ATP binds to the high-fluorescence state E_1 with high affinity of $K_1 = \sim 2$ μM, and to the low-fluorescence state E_2 with intermediate affinity of $K_2 = 60$ μM. These results show the following: a) the Ca^{2+}-free enzyme exists in at least two conformational states, high- and low-NBD fluorescence states, in both the presence and absence of ATP; b) the low-fluorescence state (E_2) has low affinity for Ca^{2+} and intermediate affinity for ATP; c) the high-fluorescence state (E_1) has high affinity for both Ca^{2+} and ATP; d) ATP accelerates Ca^{2+} binding by accelerating the conformational transition ($E_1 \rightarrow E_2$) in the Ca^{2+}-free enzyme. Such effect of ATP is caused by its binding at a single site on the enzyme (the catalytic site). Thus the different affinities of the two enzyme states for

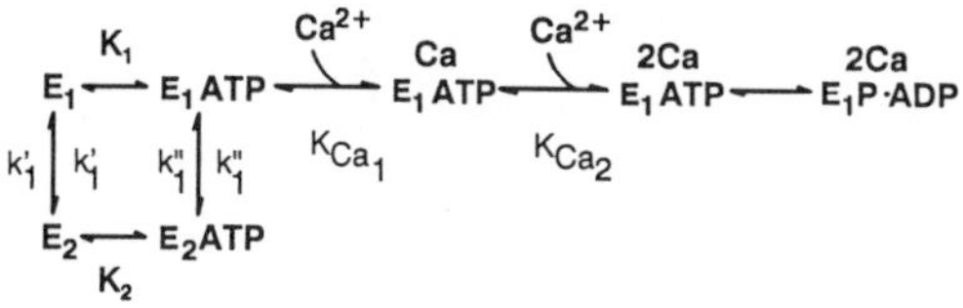

K_1	2 μM	k'_1	0.3 s^{-1}
K_2	60 μM	k'_{-1}	20 s^{-1}
K_{Ca_1}	4 μM	k''_1	3.5 ~4.3 s^{-1}
K_{Ca_2}	0.1 μM	k''_{-1}	9 s^{-1}

Fig. 6. Ca^{2+} binding to Ca^{2+}-ATPase in the presence of ATP. In this figure, Ca^{2+} binding to ATP-unbound E_1 is not shown for the sake of simplicity.

ATP provide a driving force for the observed acceleration of the conformational transition.

The results described above strongly suggest that at a physiological level of ATP concentration, binding of Ca^{2+} and ATP to the Ca^{2+}-ATPase is ordered. ATP binds first to the catalytic site of the Ca^{2+}-free enzyme and induces a conformational change, thus opening up the high-affinity Ca^{2+} site. Ca^{2+} ion then binds sequentially to form the active tertiary complex, the enzyme-ATP-Ca^{2+}, which is converted to the phosphoenzyme intermediate.

IV. EFFECTS OF H^+ AND Mg^{2+} ON Ca^{2+} BINDING TO Ca^{2+}-ATPase

Binding of Ca^{2+} to the high-affinity Ca^{2+} site of Ca^{2+}-ATPase is known to be strongly affected by pH and Mg^{2+} (*13, 26, 35–37*). H^+ and Mg^{2+} can influence Ca^{2+} binding through the following two mechanisms: one is direct competition between Ca^{2+} and these ligands for the Ca^{2+} site, and the other is through their effects on the equilibrium between the different conformational states of the free enzyme whose Ca^{2+} site has different affinities for Ca^{2+}.

The importance of the second mechanism can be illustrated by the result shown in Fig. 7 which indicates that a pH increase in the medium markedly accelerates Ca^{2+} binding and the NBD fluorescence rise associated with it. These results suggest that pH jump increases the rate of Ca^{2+} binding by accelerating the enzyme conformational change. Such conclusion is consistent with the previous observation (*16*) that the extent of the fluorescence change of the fluorescein-labeled enzyme induced by addition of Ca^{2+} or vanadate was highly dependent on the pH of the medium.

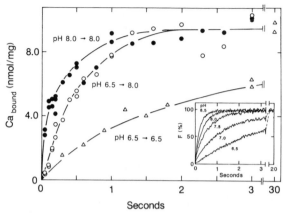

Fig. 7. Effects of pH jump on the rates of Ca^{2+} binding to the NBD-modified enzyme and the associated NBD fluorescence rise. The enzyme was preincubated with EGTA at pH 6.5 or 8.0 and then $^{45}Ca^{2+}$ binding was measured at the same or elevated pH in the presence of 50 μM Ca^{2+}. The inset shows typical traces for NBD fluorescence rise which was induced by pH jump from 6.5 to indicated pHs in the presence of 50 μM Ca^{2+}. Conditions: 0.1 M KCl, 5 mM $MgCl_2$, 11°C (for experimental details, see ref. *38*).

Mg^{2+} has been shown to induce not only a small shift in the emission spectrum of the intrinsic protein fluorescence (*37*), but also enhancement of NBD fluorescence in the Ca^{2+}-ATPase (*38*). Equilibrium fluorescence measurements showed that Mg^{2+} binding occurs at a single site with K_d values of 5–11 mM at pH 7.0 (*37*, *38*). This site seems to be Mg^{2+}-specific. The transient kinetic analysis of the Mg^{2+} effect on the pH-induced NBD fluorescence change gave results which show that Mg^{2+} stabilizes the high-fluorescence state (E_1) of the free enzyme, thus decreasing the rate of conversion from the high (E_1) to the low-fluorescence (E_2) states (*38*). This conclusion agrees well with the previous findings (*26*, *39*) that high Mg^{2+} stabilizes a form of the free enzyme unfit for phosphorylation by P_i, and that preincubation of the enzyme with high Mg^{2+} accelerates the Ca^{2+}-induced rise of intrinsic protein fluorescence.

As discussed above, the conformational transition in the Ca^{2+}-free enzyme is influenced greatly by ATP, H^+, and Mg^{2+}. An important question is how the effects of these ligands can be correlated. We recently addressed this question by using NBD fluorescence as a conformational probe (*38*). Figure 8 shows equilibrium levels of the NBD fluorescence measured at different pHs in the presence of EGTA and under various ligand conditions. In the absence of ATP or Mg^{2+}, the fluorescence level,

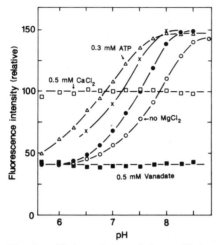

Fig. 8. pH dependences of the equilibrium level of NBD fluorescence measured under various ligand conditions. Open circles show the result of an experiment in which $MgCl_2$ was omitted from the reaction mixture. Crosses represent the fluorescence intensity at infinite Mg^{2+} concentration, which was estimated at each pH as described in the text. Experimental conditions: 0.1 M KCl, 5 mM $MgCl_2$, 0.3 mM EGTA, 11°C (for details, see ref. *38*).

which at acidic pH (6.0–6.5) was almost equal to that obtained in the presence of vanadate, increased with increasing pH, reaching the maximum at about pH 8.5. ATP at 0.3 mM (plus 5 mM $MgCl_2$) shifted the pH dependence of the equilibrium level of NBD fluorescence by about 1 pH unit toward the acidic side without influencing either the minimum or the maximum (Fig. 8). The extent of this shift was similar (about 1 pH unit) when the ATP concentration was raised from 0.3 to 1.0 mM in the presence of either 1 or 5 mM $MgCl_2$. In the presence of 5 mM $MgCl_2$ alone, the pH curve was complex; Mg^{2+} exerted a more pronounced effect at alkaline pH than at acidic pH. This seemingly uneven effect of 5 mM $MgCl_2$ was probably due to the pH-dependent change in the affinity of the enzyme for Mg^{2+}. In fact, when the NBD fluorescence intensity at the infinite Mg^{2+} concentration, which was estimated at each pH from the linear double-reciprocal plot of the observed fluorescence level *versus* the added Mg^{2+} concentration, was plotted against pH, the uneven pH dependence disappeared (Fig. 8).

These data show that at neutral or acidic pH, the MgATP- or Mg^{2+}-induced transition from low-(E_2) to high-fluorescence (E_1) states is incomplete even when a saturating concentration of each ligand is present in the reaction medium. Thus we concluded that the equilibrium between the low- (E_2) and high-fluorescence (E_1) states in the Ca^{2+}-free enzyme is primarily determined by the extent to which the enzyme is protonated and that ATP or Mg^{2+} plays a modulatory role in this enzyme-proton interaction. Analysis of the transient kinetics of the pH jump-induced NBD fluorescence change in the presence and absence of ATP and/or Mg^{2+} suggested that 1 mol each of H^+ is liberated before and after the E_2 to E_1 conformational transition, and that MgATP or Mg^{2+} promotes this conformational transition by enhancing deprotonation of the enzyme (*38*).

Figure 8 also shows that the high-fluorescence state (E_1) exhibits a fluorescence level about 1.3-fold that for the enzyme with bound calcium and about 2.6-fold that for the low-fluorescence state (E_2). This Ca^{2+}-free, high-fluorescence state is clearly the third state of the enzyme detected by the NBD fluorescence (*cf.* Fig. 4), which appears transiently only during Ca^{2+} binding to the unphosphorylated enzyme. Simulation of the NBD fluorescence change accompanying Ca^{2+} binding suggests that NBD fluorescence drops rapidly when the first Ca^{2+} ion binds to the high-fluorescence state (E_1) of the Ca^{2+}-free enzyme (upon formation of E_1Ca in scheme III of Fig. 3).

V. MECHANISM OF Ca^{2+} BINDING

Recent data (*4–7*) suggest that the Ca^{2+} site is located within a protein crevice in the membrane. Our results described in the preceding paragraphs suggest that the high affinity Ca^{2+} site becomes accessible from the cytoplasmic surface only after the unphosphorylated Ca^{2+}-ATPase undergoes conformational transition from the low-NBD fluorescence to the high-NBD fluorescence state. Deprotonation seems to play the primary role in this conformational transition and ATP or Mg^{2+} promotes the conformational transition by enhancing the deprotonation.

As already mentioned, reversal of the NBD fluorescence change in the unphosphorylated enzyme occurs in the phosphorylated enzyme when the ADP-sensitive phosphoenzyme (E_1P) is converted into the ADP-insensitive one (E_2P) (Fig. 4). The $E_1P \rightarrow E_2P$ conversion is known to be accompanied by exposure of the Ca^{2+} site to the luminal surface and by reduction of the affinity of the Ca^{2+} site for the transported Ca^{2+} ions. At present, however, we are not certain as to whether the change in the Ca^{2+} affinity of the Ca^{2+} site in the unphosphorylated enzyme is also accompanied by the change in the sidedness of the Ca^{2+} site. At any rate, the NBD label incorporated into Cys-344 in the cytosolic portion of the Ca^{2+}-ATPase appears to be able to faithfully monitor changes in the Ca^{2+} affinity of the Ca^{2+} site that is located probably within the membrane crevice. Cys-344 is located on the polypeptide segment spanning from the fourth transmembrane helix, which presumably provides an amino acid side chain to form one of the ligands for the Ca^{2+} site (6), to the region surrounding the phosphorylation site. This segment is highly conserved in many cation transport ATPases (4), which suggests the functional importance of this portion of the enzyme. This portion seems to take two major conformations during the reaction cycle, as reflected by the change in the NBD fluorescence, and may thus be important for the long-range transmission of a structural message from the Ca^{2+} binding domain to the phosphorylation domain of the Ca^{2+}-ATPase and *vice versa*.

Previous kinetical studies (25, 28) of the binding to and release of two Ca^{2+} ions from the high-affinity Ca^{2+} site in the unphosphorylated Ca^{2+}-ATPase revealed that two binding sites in the Ca^{2+}-site were apparently interacting and nonequivalent. The kinetics of exchange of $^{45}Ca^{2+}$ ions bound at the Ca^{2+} site with unlabeled Ca^{2+} ions in the reaction medium is biphasic, with a 1 : 1 stoichiometry for rapidly exchangeable to slowly exchangeable $^{45}Ca^{2+}$ ions. The rate of release of the slowly exchangeable ^{45}Ca ion decreases as the unlabeled Ca^{2+} concentration in the medium increases. In contrast, dissociation of $^{45}Ca^{2+}$ ions becomes almost monophasic when it is measured in the EGTA-containing medium. These results indicate that dissociation of two bound $^{45}Ca^{2+}$ ions from the Ca^{2+} site is ordered; release from one site is blocked by occupancy of the other. These kinetic features can be explained by a nonequivalent-site model for the high-affinity Ca^{2+} site in which dissociation of the first Ca^{2+} ion from one of the nonequivalent sites triggers a conformational change which enables the second Ca^{2+} ions to dissociate. Alternatively, these features can be explained by a model in which the high-affinity Ca^{2+} site is a binding pocket in a protein crevice in the membrane with a narrow opening toward the cytoplasm (40). In the latter model, two Ca^{2+} ions are positioned within the binding pocket in such a way that one ion sterically blocks the release of the other. Thus the overall rate of Ca^{2+} dissociation from the high-affinity Ca^{2+} site may be limited by a single rate at which Ca^{2+} leaks into the cytoplasm through the opening of the binding pocket.

SUMMARY

In this brief review, we discussed the kinetic mechanism for activation of the unphosphorylated Ca^{2+}-ATPase of sarcoplasmic reticulum by Ca^{2+} and ATP. We used fluorescence of NBD label, which was introduced into a specific sulfhydryl residue (Cys-344) near the phosphorylation site of the enzyme, as a probe to study the rapid conformational transition associated with binding to and dissociation of Ca^{2+} from the Ca^{2+} site presumably located in the membrane. According to our results, the high affinity Ca^{2+} site becomes available on the cytoplasmic surface only after the Ca^{2+}-ATPase undergoes conformational transition from the low-NBD fluorescence to the high-NBD fluorescence state. Deprotonation of functional residue(s) seems to play the primary role in this conformational transition. MgATP, by binding to the catalytic site of the low-fluorescence state, whose affinity for MgATP is about 30-fold less than that of the high-NBD fluorescence state, accelerates the conformational transition by enhancing the deprotonation. The high-NBD fluorescence state subsequently binds two Ca^{2+} ions in a sequential manner. At the physiological ATP concentration, MgATP binding occurs first and Ca^{2+} binding then follows to form the enzyme-MgATP-$2Ca^{2+}$ complex, which is converted to the phosphoenzyme.

REFERENCES

1. de Meis, L. and Vianna, A.L. *Annu. Rev. Biochem.*, **48**, 275 (1979).
2. Martonosi, A. and Beeler, T.J. *Handb. Physiol.*, **10**, S417 (1983).
3. Andersen, J.P. *Biochim. Biophys. Acta*, **988**, 47 (1989).
4. Green, N.M. and Maclennan, D.H. *Biochem. Soc. Trans.*, **17**, 819 (1989).
5. Martonosi, A.N., Jona, I., Molnar, E., Seidler, N.W., Buchet, R., and Varga, S. *FEBS Lett.*, **268**, 365 (1990).
6. Clarke, D.M., Loo, T.W., Inesi, G., and Maclennan, D.H. *Nature*, **339**, 476 (1989).
7. Inesi, G., Sumbilla, C., and Kirtley, M.E. *Physiol. Rev.*, **70**, 749 (1990).
8. Jencks, W.P. *J. Biol. Chem.*, **264**, 18855 (1989).
9. Vianna, A.L. *Biochim. Biophys. Acta*, **410**, 389 (1975).
10. Shigekawa, M., Wakabayashi, S., and Nakamura, H. *J. Biol. Chem.*, **258**, 14157 (1983).
11. Ogurusu, T., Wakabayashi, S., and Shigekawa, M. *J. Biochem.*, **109**, 472 (1991).
12. Chevallier, J. and Butow, R. *Biochemistry*, **10**, 2733 (1971).
13. Meissner, G. *Biochim. Biophys. Acta*, **298**, 906 (1973).
14. Inesi, G., Kurzmack, M., Coan, C., and Lewis, D.E. *J. Biol. Chem.*, **255**, 3025 (1980).
15. Dupont, Y. and Leigh, J.B. *Nature*, **273**, 396 (1978).

16. Pick, U. and Karlish, S.J.D. *J. Biol. Chem.*, **257**, 6120 (1982).
17. Imamura, Y., Saito, K., and Kawakita, M. *J. Biochem.*, **95**, 1305 (1984).
18. Andersen, J.P., Jorgensen, P.L., and Moller, J.V. *Proc. Natl. Acad. Sci. U.S.A.*, **82**, 4573 (1985).
19. Ikemoto, N., Morgan, J.F., and Yamada, S. *J. Biol. Chem.*, **253**, 8027 (1978).
20. Murphy, A.J. *J. Biol. Chem.*, **253**, 385 (1978).
21. Villalain, J., Gomez-Fernandez, J.C., Jackson, M., and Chapman, D. *Biochim. Biophys. Acta*, **978**, 305 (1989).
22. Buchet, R., Carrier, D., Wong, P.T.T., Jona, I., and Martonosi, A. *Biochim. Biophys. Acta*, **1023**, 107 (1990).
23. Birmachu, W. and Thomas, D.D. *Biochemistry*, **28**, 3940 (1989).
24. Guillain, F., Gingold, M.P., Büschlen, S., and Champeil, P. *J. Biol. Chem.*, **255**, 2072 (1980).
25. Dupont, Y. *Biochim. Biophys. Acta*, **688**, 75 (1982).
26. Champeil, P., Gingold, M.P., Guillain, F., and Inesi, G. *J. Biol. Chem.*, **258**, 4453 (1983).
27. Fernandez-Belda, F., Kurzmack, M., and Inesi, G. *J. Biol. Chem.*, **259**, 9687 (1984).
28. Petithory, J.R. and Jencks, W.P. *Biochemistry*, **27**, 5553 (1988).
29. Petithory, J.R. and Jencks, W.P. *Biochemistry*, **27**, 8626 (1988).
30. Inesi, G. *J. Biol. Chem.*, **262**, 16338 (1987).
31. Wakabayashi, S., Imagawa, T., and Shigekawa, M. *J. Biochem.*, **107**, 563 (1990).
32. Wakabayashi, S. and Shigekawa, M. *Biochemistry*, **29**, 7309 (1990).
33. Champeil, P., Le Maire, M., Moller, J.V., Riollet, S., Guillain, F., and Green, N.M. *FEBS Lett.*, **206**, 93 (1986).
34. Dupont, Y. *Eur. J. Biochem.*, **109**, 231 (1980).
35. Dupont, Y., Bennet, N., and Lacapere, J.J. *Ann. N.Y. Acad. Sci.*, **402**, 569 (1982).
36. Watanabe, T., Lewis, D., Nakamoto, R., Kurzmack, M., Fronticelli, C., and Inesi, G. *Biochemistry*, **20**, 6617 (1981).
37. Guillain, F., Gingold, M.P., and Champeil, P. *J. Biol. Chem.*, **257**, 7366 (1982).
38. Wakabayashi, S., Ogurusu, T., and Shigekawa, M. *Biochemistry*, **29**, 10613 (1990).
39. Loomis, C.R., Martin, D.W., McCaslin, D.R., and Tanford, C. *Biochemistry*, **21**, 151 (1982).
40. Forbush, B., III. *J. Biol. Chem.*, **262**, 11116 (1987).

Characterization of the Dihydropyridine-Sensitive, Voltage-Dependent Calcium Channel from Porcine Cardiac Sarcolemma

HANNELORE HAASE,[*1] ROLAND VETTER,[*1]
JÖRG STRIESSNIG,[*2] MARTIN HOLTZHAUER,[*1] AND
HARTMUT GLOSSMANN[*2]

*Central Institute for Cardiovascular Research, Berlin Buch, Germany[*1] and Institute of Biochemical Pharmacology, Innsbruck, Austria[*2]*

Voltage-dependent Ca^{2+} channels in the sarcolemmal membrane represent the major pathway through which Ca^{2+} enters the myocardial cell during the excitation-contraction coupling process (*1*). These channels are modulated by phosphorylation and GTP-binding proteins (for reviews see refs. *2–4*). A variety of clinically important cardioactive drugs including derivatives of 1,4-dihydropyridines, phenylalkylamines, and benzothiazepines (for reviews see refs. *5* and *6*) are known as specific pharmacological Ca^{2+} channel modulators. The 1,4-dihydropyridines have been used as high-affinity probes for the biochemical isolation of the Ca^{2+} channel protein(s) from various tissues such as skeletal and cardiac muscle. Skeletal muscle transversal (t)-tubular membranes are known as the richest mammalian source of Ca^{2+} channels/1,4-dihydropyridine receptor sites (*7*). Studies with mainly this type of membranes allowed the characterization of purified functional Ca^{2+} channels as hetero-oligomers of at least four non-covalently associated polypeptides, termed alpha$_1$ (α_1), alpha$_2$/delta (α_2/δ), beta (β), and gamma (γ) (*7, 8*). Cloning of the cDNAs of the α_1-subunits from skeletal and heart muscle and their functional expression revealed that the α_1-subunits of both tissues form a hydrophobic, membrane spanning polypeptide that effectively functions as a voltage-dependent, 1,4-dihydropyridine-sensitive Ca^{2+} channel (*9–11*). The other subunits may have stabilizing (*11*) or regulatory (*12, 13*) functions with respect to the α_1-subunit. In contrast to the skeletal

169

muscle channel much less progress has been made towards the biochemical characterization of the cardiac Ca^{2+} channel. This is mainly due to a 50 to 100-fold lower density of 1,4-dihydropyridine receptors in isolated cardiac sarcolemmal membranes compared to skeletal muscle t-tubular membrane preparation (14). The procedures for the purification of the cardiac Ca^{2+} channel protein(s) reported so far (15, 16) are very time-consuming and result in low yields of 1,4-dihydropyridine-prelabeled receptor/Ca^{2+} channel complex. In this paper, we present a new protocol for the isolation of the cardiac 1,4-dihydropyridine receptor/Ca^{2+} channel. The procedure employs as a first step a membrane isolation procedure which results in a high density of membrane-bound 1,4-dihydropyridine receptors. The 1,4-dihydropyridine receptors of these membranes are then further enriched by a three-step purification procedure which for the first time employs heparin-sepharose chromatography. The protocol was designed to allow rapid and high yield purification of cardiac 1,4-dihydropyridine receptors for further biochemical characterization of the mammalian cardiac Ca^{2+} channel protein(s).

I. ENRICHMENT OF THE MEMBRANE-BOUND 1,4-DIHYDROPYRIDINE RECEPTOR IN ISOLATED SARCOLEMMA

High and low affinity receptor sites for Ca^{2+} antagonists in cardiac membranes have been described (17–19). High affinity 1,4-dihydropyridine receptor sites were found to copurify with well established sarcolemma markers such as Na^+, K^+-ATPase, Na/Ca exchange activities (20) and β-adrenergic receptors (21) during the isolation of the sarcolemmal fraction used in the present work. Sarcolemma membranes isolated from porcine left ventricular tissue contained a single population of high affinity (K_D in the subnanomolar range) 1,4-dihydropyridine receptors as revealed by reversible binding of [^{3}H]-labeled PN200-110, nitrendipine, and azidopine (Table I). The density found of about 2.5 pmol 1,4-dihydropyridine receptors per mg of sarcolemma protein exceeds those in tissue homogenates 50 to 100-fold. Additional low affinity and high capacity 1,4-dihydropyridine receptor sites which can be detected in less purified membrane preparations became separated during the sarcolemma isolation procedure. Therefore, the highly purified cardiac sarcolemma preparations provide an excellent tool for the isolation of the cardiac Ca^{2+} channel protein.

TABLE I

1,4-Dihydropyridine Binding Parameters in Porcine Cardiac Sarcolemma

1,4-Dihydropyridine	K_D (nM)	B_{max} (pmol/mg of protein)
[³H]PN 200-110	0.16 ± 0.05	2.5 ± 0.3
[³H]nitrendipine	0.44 ± 0.10	2.4 ± 0.2
[³H]azidopine	0.25 ± 0.05	2.0 ± 0.4

Values are means $\pm$ S.E. for 4–8 different membrane preparations. [³H]1,4-dihydropyridine binding was measured at 25°C and 7.4 for 1 hr, using total ligand concentrations of 0.05 to 2.0 nM. The nonspecific binding was defined in the presence of 10^{-6} M unlabeled nitrendipine and the equilibrium binding parameters were calculated from Scatchard transformations.

II. CHARACTERIZATION OF THE MEMBRANE-BOUND RECEPTOR

A photoaffinity probe [³H]azidopine (22, 23) was employed for the identification of the 1,4-dihydropyridine receptor carrying subunit of the Ca^{2+} channel in the sarcolemmal membrane preparations. [³H]azidopine was photoincorporated covalently into 187 ± 4, 98 ± 3, 49 ± 2, and 31 ± 2 kDa ($n = 12$) proteins as determined by sodium dodecylsulfate-polyacrylamide gel electrophoresis (SDS-PAGE) under reducing conditions (Fig. 1). Photolabeling of the 187 kDa band was reduced in the presence of 1 nM and 10 nM unlabeled PN200–110 to about 50% and 100%, respectively. Moreover, micromolar concentrations of the phenylalkylamine derivative ($-$)-desmethoxyverapamil were also able to prevent the photolabeling of this protein band (data not shown). Neither the unlabeled 1,4-dihydropyridine nor the phenylalkylamine derivative protected the other polypeptides of lower M_r from labeling. These results indicate that the high M_r polypeptide of approx. 190 kDa is the Ca^{2+} channel linked 1,4-dihydropyridine receptor.

It has been reported recently that the polyanion heparin binds to an extracellular domain of the dihydropyridine sensitive Ca^{2+} channel and increases channel activity (24). Figure 2 demonstrates binding inhibition experiments using ($+$)-[³H]PN200-110 and heparin. As shown, heparin inhibited almost completely the 1,4-dihydropyridine binding to the sarcolemmal receptor. The IC$_{50}$ value for heparin of 0.15 ± 0.08 μM ($n = 3$) is similar to that obtained with the 1,4-dihydropyridine nifedipine (0.05 μM). The interaction between heparin and the 1,4-dihydropyridine receptor prompted us to use heparin sepharose for the purification of the cardiac Ca^{2+} channel.

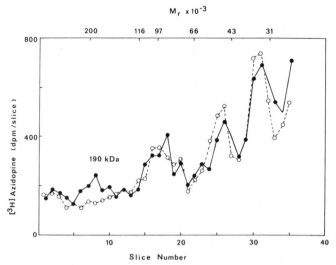

Fig. 1. Photoaffinity labeling of the 1,4-dihydropyridine receptors in the sarcolemma membrane. Sarcolemma membranes (1 mg) were incubated in the darkness
for 60 min at 25°C in 5 ml 20 mM Tris/HCl buffer, pH 7.4, containing 0.1 mM
phenylmethylsulfonyl fluoride (PMSF) and 1 nM [³H]azidopine in the absence (●)
or in the presence of 10 nM unlabeled (+)PN200-110 (○) for 1 hr. Samples were
exposed to an ultraviolet light source. The $100,000 \times g$ pellet of the photolyzed
samples was resuspended in SDS-sample buffer (5% SDS, 62.5 mM Tris/HCl, pH
7.4, 0.1 mM PMSF, 80 mM urea, 125 mM sucrose, 40 mM β-mercaptoethanol),
heated for 15 min at 56°C, and separated on a 5–15% SDS-PAGE system. Lanes
of Coomassie blue stained gels were cut into 3 mm slices, digested and counted for
[³H]radioactivity (for more technical details see refs. *22* and *23*).

III. PRELABELING AND SOLUBILIZATION

The membrane-bound 1,4-dihydropyridine receptor was labeled
with [³H]PN200-110 prior to its solubilization (Fig. 3). Prelabeling of the
cardiac receptors was a necessary step, because the affinity of the Ca^{2+}
channel receptor for the 1,4-dihydropyridine was found to decrease
markedly upon solubilization. A K_D-value of 0.34 ± 0.12 nM was determined if digitonin-solubilized sarcolemma was used in equilibrium binding studies with [³H]PN200-110. The maximal number of receptor sites
was 1.1 ± 0.2 pmol per mg of protein. This indicates that digitonin-solubilized membranes bind approx. 50% of the 1,4-dihydropyridine derivative compared to native sarcolemmal membranes (see also Table I). A
further decrease of binding affinity occurred upon purification of the
receptor.

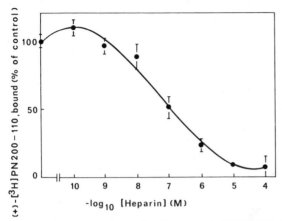

Fig. 2. Inhibition of specific binding of [³H]PN200-110 to porcine cardiac sarcolemma by heparin. Sarcolemmal membranes (32 μg of protein) were incubated with 0.5 nM (+)-[³H]PN200-110 in the presence of increasing concentrations of heparin (molecular weight 7,500-15,000) under standard conditions for 1,4-dihydropyridine binding (see Table I). The maximum molecular weight value was used to calculate the heparin concentration. Specific binding of (+)-[³H]-PN200-110 in the absence of heparin (control) was 1.20 ± 0.06 pmol/mg. Values are means $\pm$ S.D. for triplicate determinations of a typical experiment.

Treatment of sarcolemmal membranes with 0.7-1.0% digitonin in the presence of a protease inhibitor mixture (see legend to Fig. 4) resulted in the solubilization of almost equal amounts of prelabeled receptor and membrane protein ($60\pm9\%$, $n=9$) leaving the specifically bound radioactivity unchanged. The prelabeled ligand-receptor complex was stable at 4°C and had a half-life of 10 ± 1 hr ($n=3$). The half-life of this complex was prolonged by the Ca²⁺ antagonist (+)-*cis* diltiazem which is a positive heterotropic allosteric modulator of 1,4-dihydropyridine-binding (*25*).

IV. PURIFICATION OF THE SOLUBILIZED 1,4-DIHYDROPYRIDINE RECEPTOR

The solubilized cardiac 1,4-dihydropyridine receptor was purified by a rapid three-step procedure consisting of chromatography on wheat germ agglutinin sepharose, sucrose density gradient centrifugation, and chromatography on heparin sepharose (Fig. 3). The procedure was completed within 12 hr and resulted in a 240-fold enrichment of the bound radiolabeled Ca²⁺ antagonist over the original detergent extract. It

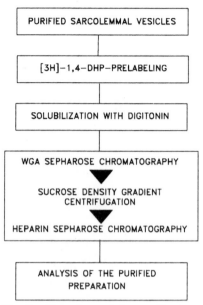

Fig. 3. Flow chart of the purification procedure for the 1,4-dihydropyridine receptor of the porcine cardiac Ca^{2+} channel. Porcine left ventricular tissue (500 g) was used for the isolation of sarcolemma membrane vesicles, yielding 20–40 mg of membrane protein. Using these membranes as starting material, 20–40 μg of 1,4-dihydropyridine receptor Ca^{2+}-channel complex could be isolated with a specific activity of 283 ± 45 pmol/mg.

was based on procedures used for the purification from skeletal muscle microsomes, *i.e.*, affinity purification on lectin sepharose followed by sucrose density gradient centrifugation (*22*). After these two purification steps the cardiac preparation was enriched about 70-fold from the starting material. Finally, the peak fractions from the sucrose density gradient centrifugation were loaded onto a heparin sepharose column (Fig. 4). The bound receptor was eluted using a linear 0.1–1 M sodium chloride gradient resulting in a further 3- to 4-fold enrichment. In addition to its properties as a weak ion exchanger, heparin sepharose may also function as an affinity column for the 1,4-dihydropyridine receptor complex which contributes to an effective purification. The final calculated specific activity was at least 590 pmol/mg of protein (240-fold enrichment from a specific activity of 2.5 pmol/mg). Thus the expected purity for our preparation is at least 22% based on the maximal theoretical specific activity of 2,700 pmol/mg of protein derived from a 370-kDa

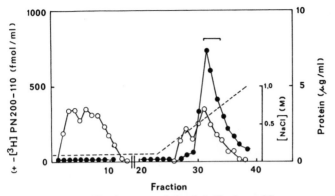

Fig. 4. Final purification step of the 1,4-dihydropyridine receptor by heparin sepharose-chromatography. Porcine cardiac sarcolemma membranes (20-40 mg) were prelabeled with (+)-[³H]PN200-110 at a ligand-to-protein ratio of 2.5 to 1 (pmol per mg) and solubilized, using 1% digitonin. The solubilized 1,4-dihydropyridine receptors were purified by lectin sepharose chromatography and sucrose density gradient centrifugation. Finally, fractions (12-16 ml) containing the highest activity were applied onto a heparin sepharose column (6 ml packed volume) equilibrated in 20 mM Tris-HCl buffer, pH 7.4, 0.1% digitonin, 0.1 M NaCl, 0.1 mM PMSF, 1 mM iodoacetamide, 0.1 mM benzamidine, 1 μM pepstatin A, and 10 μM (+)-*cis* diltiazem (buffer A). The gel was washed with 50 ml of buffer A. The receptor was then eluted with 30 ml of a linear 0.1-1 M NaCl gradient in buffer A. Profiles of receptor (●), protein (○), and NaCl (broken line) concentrations are given. The concentrations of prelabeled receptors and protein were measured by quantitation of protein-bound (+)-[³H]PN200-110 using the polyethylene glycol precipitation technique (*25*).

protein (determined for rat cardiac 1,4-dihydropyridine receptors) (*26*)). Figure 5 exemplifies the polypeptide composition of the preparation after each purification step in silver-stained SDS gels.

V. CHARACTERIZATION OF THE PURIFIED RECEPTOR

Analysis of the purified porcine cardiac receptor by SDS-PAGE revealed a major polypeptide at 180-200 kDa if analyzed under alkylating conditions (Fig. 5, lane 6). This band represents approximately 40% of the total silver staining intensity. After reduction of disulfide bonds the broad diffuse silver stained band at the 180-200 kDa region was resolved into a 150 kDa and a 190 kDa band (Fig. 5, lanes 4 and 6: Fig. 6, inset). This suggests that the 180-200 kDa band consists of both a reduction-insensitive (termed α_1 in analogy to skeletal muscle (*23*)) and a reduction-sensitive component (termed α_2 in analogy to skeletal muscle (*23*)). A

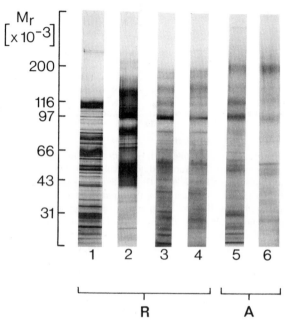

Fig. 5. Polypeptide pattern of the cardiac 1,4-dihydropyridine receptor at each step of purification. Samples were denatured either in the presence of 10 mM dithiothreitol (reducing conditions, R) or 10 mM N-ethylmaleimide (alkylating conditions, A), separated by SDS-PAGE using 5–15% gradient gels and stained with silver: Lane 1, digitonin-solubilized sarcolemma (9 µg); lane 2, wheat germ agglutinin (WGA) purified fraction (9 µg); lanes 3 and 5, sucrose gradient pool (5 µg); lanes 4 and 6, purified material after heparin sepharose chromatography (3 µg); (protein given in parenthesis).

comparison of purified skeletal muscle and the cardiac preparations revealed a higher apparent molecular mass for the cardiac α_1 subunit (190 kDa *vs.* 170 kDa) in agreement with photoaffinity labeling experiments (Fig. 6). The 98 kDa polypeptide appeared to be a loosely attached impurity as its relative staining intensity decreased relative to the α_1 subunit after the heparin sepharose step (compare lanes 5 and 6 in Fig. 5). Although polypeptides around 50 and 30 kDa were further enriched in this step, their unequivocal identification as copurifying subunits of the cardiac channel (equivalent to the skeletal muscle β and γ subunits) was impossible. The polypeptide composition as well as the purity is comparable to cardiac Ca^{2+} antagonist receptors purified from bovine (*15*) or

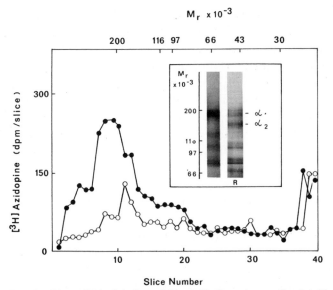

Fig. 6.   Photoaffinity labeling of the purified porcine cardiac 1,4-dihydropyridine receptor. Sarcolemma membranes were prelabeled with 10 nM (−)-[³H]azidopine, solubilized and purified by the three-step protocol. Covalent incorporation of (−)-[³H]azidopine into proteins was achieved by UV irradiation of the WGA-purified material for 30 min with a Philips 38W/TL blacklight lamp at a distance of 10 cm on ice. After final purification the receptor was concentrated by lyophilization after dialysis against 0.05% SDS at room temperature. Samples of 5 μg protein were subsequently subjected to SDS-PAGE under alkylating (●) and reducing (○) conditions using 10 mM N-ethylmaleimide and 10 mM dithiothreitol, respectively. After electrophoresis the gel was silver stained and further processed as described in the legend to Fig. 1. Inset: silver stained gel in the high M_r region under alkylating (A) and reducing (R) conditions.

chicken heart (*16*) with more complicated protocols (five-step purifications). The yields of prelabeled receptors are higher (20%, compared to 2.5% (*15*) and 8% (*16*)) and the entire procedure can be completed within 12 hr after solubilization (as compared to 24–48 hr) due to the lower number of purification steps.

Figure 6 shows the results of photoaffinity labeling experiments of the cardiac α_1 subunit after prelabeling with (−)-[³H]azidopine. The incorporated radioactivity comigrated exactly with the silver-staining intensity of the cardiac 180–200 kDa polypeptide under both alkylating and reducing SDS-PAGE conditions. Photoaffinity labeling was specific as inclusion of 200 nM (+)-PN200-110 completely suppressed incorporation into the α_1-subunit (data not shown). In comparison to the photoaffinity labeling experiments with intact sarcolemma (Fig. 1) the results suggest that the cardiac α_1 subunit did not undergo detectable proteolytic degradation upon receptor purification. We noted, however, in both sarcolemma and purified receptor preparation, traces (approx. 20% of the main radioactivity peak) of specifically photoincorporated label around 250 kDa, perhaps representing another α_1-subunit. In accordance with the 1,4-dihydropyridine receptor from bovine and chicken heart (*15, 16*), the porcine cardiac α_1 subunit migrated at a higher apparent molecular mass upon SDS-PAGE than α_1 of the skeletal muscle as shown by photolabeling.

VI. PHOSPHORYLATION WITH CYCLIC AMP-DEPENDENT PROTEIN KINASE

Cyclic AMP-dependent phosphorylation is an established principle of Ca^{2+} channel regulation in the myocardium (*1, 4*). However, in contrast to the skeletal muscle Ca^{2+} channel (*27–31*) none of the porcine cardiac α subunits or the other polypeptides present in the preparation underwent rapid and substantial phosphorylation by the cyclic AMP-dependent protein kinase (catalytic subunit).

In agreement with these phosphorylation studies, the amino acid sequences deduced from cloned rabbit skeletal muscle and cardiac α_1 subunits (*9–11, 32*) suggest that the cardiac protein lacks the phosphorylation site between the membrane spanning domains II and III, which is rapidly phosphorylated by the kinase *in vitro* (*33*). Thus, a previously generally accepted physiologic regulation (*e.g.*, activation) *via* receptors and the cAMP-cascade in heart cannot yet be confirmed with the detergent-purified channel complex from either avian (*16*) or mammalian tissue.

Acknowledgments

This work was supported by grants to H.G. from Fonds zur Förderung der wissenschaftlichen Forschung (FWF) and Österreichische Nationalbank.

REFERENCES

1. Reuter, H. *Nature*, **301**, 569 (1983).
2. Tsien, R.W., Bean, B.P., Hess, P., Lansmann, J.B., Nilius, B., and Novycky, M.C. *J. Mol. Cell Cardiol.*, **18**, 691 (1986).
3. Trautwein, W., Cavalie, A., Flockerzi, V., Hofmann, F., and Pelzer, D. *Circulation Res.*, **61** (Suppl. I), 1 (1987).

4. Trautwein, W. and Hescheler, J. *Annu. Rev. Physiol.*, **52**, 257 (1990).
5. Fleckenstein, A. *Annu. Rev. Pharmacol. Toxicol.*, **17**, 149 (1977).
6. Glossmann, H. and Striessnig, J. *Vitam. Horm.*, **44**, 155 (1988).
7. Glossmann, H. and Striessnig, J. *Rev. Physiol. Biochem. Pharmacol.*, **114**, 1 (1990).
8. Catterall, W.A., Seagar, M.J., and Takahashi, M. *J. Biol. Chem.*, **263**, 3535 (1989).
9. Tanabe, T., Takeshima, H., Mikami, A., Flockerzi, V., Takahashi, H., Kangawa, K., Kojima, M., Matsuo, H., Hirose, T., and Numa, S. *Nature*, **328**, 313 (1987).
10. Perez-Reyes, E., Kim, H.S., Lacerda, A.E., Horne, W., Wei, X., Rampe, D., Campbell, K.P., Brown, A.M., and Birnbaumer, L. *Nature*, **340**, 233 (1989).
11. Mikami, A., Imoto, K., Tanabe, T., Niidome, T., Mori, Y., Takeshima, H., Narumiya, S., and Numa, S. *Nature*, **340**, 231 (1989).
12. Ruth, P., Röhrkasten, A., Biel, M., Bosse, E., Regulla, S., Meyer, H.E., Flockerzi, V., and Hofmann, F. *Science*, **245**, 1115 (1989).
13. Nunoki, V., Nunoki, K., Florio, V., and Catterall, W.A. *Proc. Natl. Acad. Sci. U.S.A.*, **86**, 6816 (1989).
14. Hofmann, F., Oeken, H.J., Schneider, T., and Sieber, M. *J. Cardiovasc. Pharmacol.*, **12** (Suppl. 1), S25 (1988).
15. Schneider, T. and Hofmann, F. *Eur. J. Biochem.*, **174**, 127 (1988).
16. Chang, C.F. and Hosey, M.M. *J. Biol. Chem.*, **263**, 18929 (1988).
17. Bellemann, P., Ferry, D., Luebecke, F., and Glossmann, H. *Drug Res.*, **31**, 2064 (1981).
18. Ptasienski, J., McMahon, K.J., and Hosey, M.M. *Biochem. Biophys. Res. Commun.*, **129**, 910 (1985).
19. Ruth, P., Flockerzi, V., von Nettelbladt, E., Oeken, J., and Hofmann, F. *Eur. J. Biochem.*, **150**, 313 (1985).
20. Haase, H., Vetter, R., Will, H., and Will-Shahab, L. *Biomed. Biochim. Acta*, **45**, S223 (1986).
21. Haase, H., Wallukat, G., Vetter, R., and Will, H. *Biomed. Biochim. Acta*, **46**, S363 (1987).
22. Striessnig, J., Moosburger, K., Goll, A., Ferry, D.R., and Glossmann, H. *Eur. J. Biochem.*, **161**, 603 (1986).
23. Vaghy, P.L., Striessnig, J., Miwa, K., Knaus, H.G., Itagaki, K., McKenna, E., Glossmann, H., and Schwartz, A. *J. Biol. Chem.*, **262**, 14337 (1987).
24. Knaus, H.G., Scheffauer, F., Romanin, C., Schindler, H.G., and Glossmann, H. *J. Biol. Chem.*, **265**, 1 (1990).
25. Glossmann, H. and Ferry, D.R. *Methods Enzymol.*, **109**, 513 (1984).
26. Horne, W.A., Eiland, G.A., and Oswald, R.E. *J. Biol. Chem.*, **261**, 3588 (1986).
27. Curtis, B.M. and Catterall, W.A. *Proc. Natl. Acad. Sci. U.S.A.*, **82**, 4255 (1985).
28. Nastainczyk, W., Röhrkasten, A., Sieber, M., Rudolph, C., Schächtele, C., Marme, D., and Hofmann, F. *Eur. J. Biochem.*, **168**, 137 (1987).
29. Hosey, M.M., Barhanin, J., Schmid, A., Vandaele, S., Ptasienski, J., O'Callahan, C., Cooper, C., and Lazdunski, M. *Biochem. Biophys. Res. Commun.*, **147**, 1137 (1987).
30. Imagawa, T., Leung, A.T., and Campbell, K.P. *J. Biol. Chem.*, **262**, 8333 (1987).
31. Hymel, L., Striessnig, J., Glossmann, H., and Schindler, H.G. *Proc. Natl. Acad. Sci. U.S.A.*, **85**, 4290 (1988).
32. Ellis, S.B., Williams, M.E., Ways, N.R., Brenner, R., Sharp, A.H., Leung, A.T., Campbell, K.P., McKenna, E., Koch, W.J., Hui, A., Schwartz, A., and Harpold, M.M. *Science*, **241**, 1661 (1988).

33. Röhrkasten, A., Meyer, H.E., Nastainczyk, W., Sieber, M., and Hofmann, F. *J. Biol. Chem.*, **263**, 1525 (1988).

Molecular Machinery of Calcium Release from Cardiac Sarcoplasmic Reticulum

MAKOTO INUI

Departments of Neurochemistry and Pathophysiology, Osaka University School of Medicine, Suita, Osaka 565, Japan

Contraction and relaxation of heart muscle are controlled by free calcium ion (Ca) in the myoplasm, which is regulated by the two membrane systems, sarcolemma/transverse tubules and sarcoplasmic reticulum (SR) (*1, 2*). Following depolarization of the sarcolemma/transverse tubule, extracellular Ca enters heart cells through a voltage-dependent Ca channel. The entering Ca then induces the release of Ca from the SR, thereby inducing contraction (*3, 4*). When Ca is reaccumulated into the SR by a Ca pump, the muscle relaxes. The Ca pump protein of SR has been extensively studied at the molecular level (*1, 5*), and significant progress has recently been made in defining the molecular machinery of the Ca release channels of SR (*2*). This was made possible by use of a plant alkaloid, ryanodine, as a specific ligand of the Ca release channel. The Ca release channel of SR is therefore called the "ryanodine receptor", as the voltage-dependent Ca channel of the sarcolemma/ transverse tubules is called the "dihydropyridine receptor". This article focuses on cardiac SR, and reviews recent advances in studies on the molecular machinery of SR Ca release and its role in excitation-contraction coupling.

I. CHARACTERIZATION OF TERMINAL CISTERNA AND LONGITUDINAL TUBULE VESICLES

Cardiac SR has two domains, terminal cisternae and longitudinal tubules (*6*). Morphological studies of the heart muscle by electron microscopy revealed that the terminal cisternae have foot structures

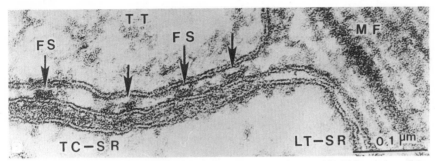

Fig. 1. Electron micrograph of heart muscle membranes (7). Thin section of dog heart muscle. Terminal cisterna of SR (TC-SR) is associated with transverse tubule (TT) by way of foot structures (FS), forming a diad junction. The compartment of the terminal cisterna is filled with electron opaque contents, in contrast with the longitudinal tubule of SR (LT-SR). MF, myofilaments.

which connect the terminal cisternae with the sarcolemma/transverse tubules, and that the terminal cisternae are filled with electron opaque contents (7) (Fig. 1). The latter have been identified as the Ca binding protein (calsequestrin) which keeps Ca within the terminal cisternae during relaxation of the heart muscle (8). The longitudinal tubules, on the other hand, consist only of Ca pump membranes and are devoid of contents in their lumen and of foot structures. These morphological characteristics of cardiac SR are essentially the same as those of skeletal muscle SR.

Functions of SR can be studied using isolated SR vesicles. Cardiac SR was subfractionated by Jones and Cala (9) using Ca-oxalate loading method into "ryanodine-sensitive" and "ryanodine-insensitive" vesicles. They found that the Ca loading of the ryanodine-sensitive vesicles was stimulated by submillimolar levels of ryanodine. We subfractionated cardiac SR by Ca-phosphate loading and subsequent density gradient centrifugation into two populations of vesicles, and determined from the morphological characteristics that one originates from longitudinal tubules and the other from terminal cisternae (7). The latter vesicles retain morphologically intact foot structures and electron opaque contents. The protein profile of subfractions shows that the main component is the Ca pump protein (about 40%) in both vesicles. Phospholamban, a regulatory protein of Ca pump of cardiac SR, is also distributed evenly in the two vesicles. The terminal cisterna vesicles are enriched in calsequestrin and have a characteristic high molecular weight protein.

To characterize the Ca release from cardiac SR, the subfractions were examined in terms of Ca permeability and its responsiveness to reagents such as ruthenium red and ryanodine. Ruthenium red was shown to close the Ca release channels of skeletal muscle SR (*10*), and ryanodine was demonstrated by Fleischer *et al.* (*11*) to modulate the Ca release of skeletal muscle SR at nanomolar concentrations but not at submillimolar concentrations. The longitudinal tubule fraction of cardiac SR has a high Ca loading rate which is insensitive to ruthenium red (*7*). In contrast, the terminal cisterna fraction has a low Ca loading rate which is enhanced about 5-fold by ruthenium red, although the Ca-dependent ATPase activity is unchanged by ruthenium red addition (*7*). When [^{3}H]ryanodine binding is measured in the subfractions of cardiac SR, ryanodine specifically binds to the terminal cisterna vesicles but not to the longitudinal tubule vesicles (*7*). These observations suggest that the Ca release channel is localized in the terminal cisternae of SR in heart muscle.

II. RYANODINE BINDING AND Ca RELEASE CHANNELS

Since the Ca release channels of cardiac SR are localized in the terminal cisternae, the Ca release was further characterized by ryanodine in the terminal cisterna vesicles. Ryanodine binding in this fraction is of two types with high ($K_D \leq 10$ nM) and low ($K_D \geq 1 \mu$M) affinities (*7*) (Table I). When the pharmacological actions of ryanodine were examined in the terminal cisterna vesicles, ryanodine blocked the action of ruthenium red to enhance the Ca loading at nanomolar concentrations (*7*). The concentration of this effect (K_i) is in the same range as the K_D for the high affinity ryanodine binding. On the other hand, high concentrations (micromolar range) of ryanodine enhanced the Ca loading (*7*). The K_m is about 1 μM (Table I). Thus, ryanodine has two different actions on the Ca release channels of the terminal cisternae of cardiac SR as reported in skeletal muscle SR (*11*). High affinity binding of ryanodine locks the channel in the open state, whereas low affinity binding closes the channel.

TABLE I
Ryanodine Binding and Modulation of the Function of Terminal Cisternae of Cardiac SR

Ryanodine binding	K_d (nM)	B_{max} (pmol/mg)	Ryanodine action on channel	K_i or K_m (nM)
High affinity site	8	5.1	Locks channel open (K_i)	7
Low affinity site	1,082	32.8	Closes channel (K_m)	1,110

These observations indicate that ryanodine is a specific ligand for the Ca release channel of the terminal cisternae of SR in heart muscle as well as skeletal muscle. The finding that ryanodine binds to the Ca release channel at nanomolar range was especially important in leading to the identification of the channel machinery and the successful purification of it.

III. PURIFICATION AND CHARACTERIZATION OF RYANODINE RECEPTOR/ Ca RELEASE CHANNEL

The ryanodine receptor has been isolated from cardiac SR by monitoring the ryanodine binding activity (12, 13). The procedure was modified from that used for skeletal muscle SR (13, 14). Cardiac microsomes were solubilized with a zwitterionic detergent (CHAPS) in the presence of phospholipids, and the receptor protein was purified by three sequential types of column chromatography. The purified receptor showed a single band of M_r 340,000 on sodium dodecyl sulfate-polyacrylamide gel electrophoresis (12). However, the molecular weight of the peptide was later found to be higher (about 550,000). When the ryanodine receptor proteins from heart and skeletal muscle were compared, the electrophoretic mobilities were slightly different between the two (12), and the antibody against the skeletal muscle receptor did not crossreact with the heart receptor, indicating that the two proteins are different polypeptides.

The purified receptor retained the same characteristics of ryanodine binding as those of the terminal cisterna vesicles, in that high and low affinity bindings were observed (12). When the purified receptor was reconstituted into lipid bilayers, the purified receptor showed the Ca channel activity (15). The channel activity was activated by submicromolar Ca, potentiated by ATP and ryanodine, and inhibited by ruthenium red and Mg^{2+}. These characteristics of Ca channel activity of the purified receptor are the same as those observed in cardiac SR vesicles (16).

Negative stain electron microscopy of the purified receptor from heart muscle revealed a 4-fold symmetric structure identical in shape and size to the foot structure (12). Thus, the ryanodine receptor is identical to the Ca release channel, and is equivalent to the foot structure which connects the terminal cisternae to the sarcolemma/transverse tubules. The structural detail of skeletal muscle receptor has been analyzed (17)

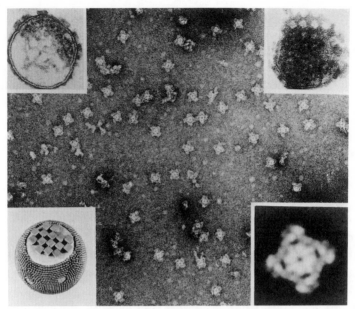

Fig. 2.   Morphology of the ryanodine receptor/Ca release channel (*17*). Electron
micrograph of negatively stained purified ryanodine receptors from skeletal muscle
(center). Essentially the same morphology is observed with purified ryanodine
receptor from heart. Upper left insert, the thin section of the terminal cisterna
vesicles; upper right insert, the section tangential to the surface of the junctional
face membrane of the terminal cisterna; lower left insert, model of the terminal
cisterna vesicle; lower right insert, computer-averaging of 240 images of the
purified receptor.

(Fig. 2). The size of the foot structure was estimated to be about 2,300,000
by scanning transmission electron microscopy (*18*), therefore this probably consists of four identical polypeptides of 550,000 daltons. Furthermore, electron image reconstruction revealed the three-dimensional architecture of the Ca release channel in that the central channel connects with
four radial channels, which appear to empty into the junctional gap
between the terminal cisternae of SR and the sarcolemma/transverse
tubules (*19*). Since morphology of the heart receptor is quite similar to
that of skeletal muscle, the two Ca release channels may have similar
architecture.

Recently, the amino acid sequence of cardiac ryanodine receptor was
deduced from cDNA by Otsu *et al.* (*20*) and Nakai *et al.* (*21*). When
compared with the skeletal muscle ryanodine receptor, the two are
recognized as being homologous in amino acid sequence although they

are different gene products (20). They exhibit the same molecular architecture in that the carboxyl-terminal hydrophobic region is the transmembrane domain and the remaining large hydrophilic portion constitutes the cytoplasmic domain (foot structure). Since the carboxyl-terminal region including the transmembrane region in the carboxyl-terminal part is highly conserved between the two receptors, this region is supposed to play an important role in forming the Ca release channel. Interestingly, this region also shows significant similarity to the corresponding region of IP_3 receptor (22).

IV.   EXCITATION-CONTRACTION COUPLING OF HEART MUSCLE

In contraction of heart muscle, extracellular Ca must first enter the heart cell through a voltage-dependent Ca channel following depolarization of the sarcolemma/transverse tubule. The newly entered Ca then induces the Ca release from the terminal cisterna of SR through a Ca release channel. This process has been referred to as "Ca-induced Ca release" (3, 4). Thus, two types of Ca channels are functionally coupled with each other, and both play important roles in this process (2). Since the Ca release channel exists as a foot structure connecting the terminal cisterna and sarcolemma/transverse tubule, the structural association of these two channels can be assumed. Ca entry through a voltage-dependent Ca channel, however, is not necessary in contraction of skeletal muscle (3). Instead, the voltage-dependent Ca channel serves as a voltage sensor in the transverse tubules of skeletal muscle (23), providing "depolarization-induced Ca release". In this process, the voltage sensor is thought to induce the conformational change of the ryanodine receptor, opening the Ca release channel; thus the mechanism of Ca release from cardiac SR is different from that of skeletal muscle SR. As mentioned above, however, the structure and function of the cardiac Ca release channel is very similar to that in skeletal muscle. Tanabe et al. (24) recently demonstrated that the cardiac type of excitation-contraction coupling was restored when the cardiac voltage-dependent Ca channel was expressed in skeletal muscle cells of mutant mice with muscular dysgenesis, which alters the skeletal muscle voltage-dependent Ca channel gene and prevents its expression. Therefore, the difference in excitation-contraction coupling may be due to the difference of the voltage-dependent Ca channels of the sarcolemma/transverse tubules. It remains to be seen whether the Ca release channels of SR directly interact with the voltage-dependent Ca channel of the sarcolemma/transverse tubules and, if so, how they interact.

SUMMARY

SR plays a central role in controlling free Ca in heart muscle cells. We found that a plant alkaloid, ryanodine, is a specific modulator of the Ca release channel of cardiac SR as well as skeletal muscle SR. The Ca release channel is localized in the terminal cisternae but not in the longitudinal tubules of cardiac SR. Ryanodine at nanomolar range locks the Ca release channel in the open state, while ryanodine closes the channel at micromolar range. Using ryanodine as a specific ligand, the ryanodine receptor was isolated and found to be equivalent to the foot structures which connect the terminal cisternae of SR to the sarcolemma/ transverse tubules. On reconstitution into bilayers, the receptor was identified as the Ca release channel of cardiac SR. The foot structure consists of four identical monomers of about 500,000 daltons, showing four-fold symmetry. The transmembrane segments of monomers are the thought to form the Ca channel at the center of the foot structure. The Ca release channel of cardiac SR is coupled with the voltage-dependent Ca channel of the sarcolemmal/transverse tubule in excitation-contraction coupling. The two Ca channels are responsible for the "Ca-induced Ca release" mechanism.

Acknowledgment
The author is grateful to Dr. Sidney Fleischer and to the many collaborators who participated in these studies.

REFERENCES

1. Tada, M., Yamamoto, T., and Tonomura, Y. *Physiol. Rev.*, **58**, 1 (1978).
2. Fleischer, S. and Inui, M. *Annu. Rev. Biophys. Biophys. Chem.*, **18**, 333 (1989).
3. Endo, M. *Physiol. Rev.*, **57**, 71 (1977).
4. Fabiato, A. *Am. J. Physiol.*, **245**, C1 (1983).
5. MacLennan, D.H., Brandl, C.J., Korczak, B., and Green, N.M. *Nature*, **316**, 696 (1985).
6. Sommer, J.R. and Jennings, R.B. *In* "The Heart and Cardiovascular System," Vol. 2, ed. H. Fozzard, R. Jennings, E. Haber, A. Katz, and H. Morgan, p. 845 (1986). Raven Press, New York.
7. Inui, M., Wang, S., Saito, A., and Fleischer, S. *J. Biol. Chem.*, **263**, 10843 (1988).
8. Jorgensen, A.O. and Campbell, K.P. *J. Cell Biol.*, **981**, 597 (1984).
9. Jones, L.R. and Cala, S.E. *J. Biol. Chem.*, **256**, 11809 (1981).

10. Ohnishi, S.T. *J. Biochem.*, **86**, 1147 (1979).
11. Fleischer, S., Ogunbunmi, E.M., Dixon, M.C., and Fleer, E.A. *Proc. Natl. Acad. Sci. U.S.A.*, **82**, 7256 (1985).
12. Inui, M., Saito, A., and Fleischer, S. *J. Biol. Chem.*, **262**, 15637 (1987).
13. Inui, M. and Fleischer, S. *Methods Enzymol.*, **159**, 490 (1988).
14. Inui, M., Saito, A., and Fleischer, S. *J. Biol. Chem.*, **262**, 1740 (1987).
15. Hymel, L., Schindler, H., Inui, M., and Fleischer, S. *Biochem. Biophys. Res. Commun.*, **152**, 308 (1988).
16. Rousseau, E. and Meissner, G. *Am. J. Physiol.*, **256**, H328 (1989).
17. Saito, A., Inui, M., Radermacher, M., Frank, J., and Fleischer, S. *J. Cell Biol.*, **107**, 211 (1988).
18. Saito, A., Inui, M., Wall, J.S., and Fleischer, S. *Biophys. J.*, **55**, 206a (1989).
19. Wagenknecht, T., Grassucci, R., Frank, J., Saito, A., Inui, M., and Fleischer, S. *Nature*, **338**, 167 (1989).
20. Otsu, K., Willard, H.F., Khanna, V.K., Zorzato, F., Green, N.M., and MacLennan, D.H. *J. Biol. Chem.*, **265**, 13472 (1990).
21. Nakai, J., Imagawa, T., Hakamata, Y., Shigekawa, M., Takeshima, H., and Numa, S. *FEBS Lett.*, **271**, 169 (1990).
22. Furuichi, T., Yoshikawa, S., Miyawaki, A., Wada, K., Maeda, N., and Mikoshiba, K. *Nature*, **342**, 32 (1989).
23. Rios, E. and Brum, G. *Nature*, **325**, 717 (1987).
24. Tanabe, T., Mikami, A., Numa, S., and Beam, K.G. *Nature*, **344**, 451 (1990).

The Molecular Regulation of Muscarinic K$^+$ Channels by GTP-Binding Proteins in Cardiac Atrial Cell Membrane

YOSHIHISA KURACHI,*[1,*2] HIROYUKI ITO,*[2]
REIKO TAKIKAWA,*[2] TOSHIAKI NAKAJIMA,*[2] AND
TSUNEAKI SUGIMOTO*[2]

Division of Cardiovascular Diseases, Department of Internal Medicine, Mayo Clinic, Mayo Foundation, Rochester, MN 55905, U.S.A.,[*1] and The 2nd Department of Internal Medicine, Faculty of Medicine, University of Tokyo, Tokyo 113, Japan*[*2]*

Activation of muscarinic acetylcholine (m-ACh) or adenosine (Ado)-receptors in the heart causes slowing of the heart rate and atrioventricular conduction by increasing K$^+$ conductance (*1-3*). The time course of activation of the macroscopic K$^+$ current by ACh shows a sigmoidal onset, which suggests a multistep process of the activation (*4, 5*). Recently it was shown that GTP-binding proteins (G), whose functions are inhibited by pertussis toxin, are involved in muscarinic activation of the K$^+$ current (*6-8*). In the present study, we examined the molecular mechanisms underlying activation and desensitization of the K$^+$ channels by G proteins in the cardiac atrial cell membrane (*9-14*).

I. PREPARATION

The atrial cells isolated from the adult guinea pig heart were used (*9*). Membrane currents and membrane potentials were recorded with glass pipettes in the "cell-attached" and "inside-out" patch recordings and whole cell clamp recording configurations (*15*). The pipette solution contained (in mM): KCl 145, CaCl$_2$ 1, MgCl$_2$ 1, HEPES (N-2-hydroxyethylpiperazine-N'-2-ethanesulfonic acid)-KOH 5, pH 7.4, and various concentrations of ACh or Ado. Cells were perfused by Tyrode solution (35-37°C) of the following composition (in mM): NaCl 136.5, KCl 5.4, CaCl$_2$ 1.8, MgCl$_2$ 0.53, glucose 5.5, HEPES-NaOH buffer 5, pH 7.4. In the inside-out patches, the bath was perfused with the internal solution (in mM): K aspartate 120, KCl 20, $MgCl_2$ 1, Na_2ATP 3, EGTA 5, HEPES-KOH buffer 5 or KCl 140, $MgCl_2$ 1, Na_2ATP 3, EGTA 5, HEPES-KOH buffer 5, pH 7.2–7.3. In the whole cell recordings, the pipette solution was (in mM): K aspartate 130, KCl 20, KH_2PO_4 1, $MgCl_2$ 1–2, EGTA 5, Na_2ATP 3, HEPES-KOH buffer 5, pH 7.2–7.3. GTP or GTP-γS 100 μM was also added to the internal solution.

Pertussis toxin, islet-activating protein (IAP), was a gift from Professor Michio Ui. For preactivation, IAP (50 μg) with dithiothreitol (5 mM) was incubated in 5 ml of the internal solution (3 mM ATP) at 37°C for 15–20 min. The solution was then diluted to 50 ml with the internal solution.

II. SINGLE CHANNEL CURRENTS OF ACETYLCHOLINE AND ADENOSINE-REGULATED K^+ CHANNELS

In 1983, Belardinelli and Isenberg showed that both ACh and Ado shorten the atrial action potential and hyperpolarize the resting membrane by increasing an inward-rectifying K^+ current (16). They proposed the hypothesis that ACh and Ado regulate the same K^+ channel in the atrial cell membrane. We compared properties of the channels activated by ACh and Ado in guinea pig atrial cells using the cell-attached patch clamp technique (Fig. 1).

Similar pulse-like channel currents were recorded with pipettes containing Ado and ACh. The frequency of the channel openings with Ado pipettes was greatly attenuated by 100 μM theophylline (P1-purinergic receptor antagonist) but was not affected by 20 μM atropine (muscarinic receptor antagonist). In contrast, the channel openings with ACh pipettes were only reduced by atropine. The amplitude of the unitary channel current at the resting level with 20 mM extracellular K^+ (Er ~ -52 mV) was -2.2 ± 0.3 pA ($n=5$) with ACh pipettes and was -2.1 ± 0.4 pA ($\tilde{n}=10$) with Ado pipettes. The unitary current increased with hyperpolarization and decreased with depolarization in both cases. Current traces became flat at around the K^+ equilibrium potential ($\sim$ Er $+52$ mV). At more positive potentials, small outward currents through the channels were sometimes observed. The current-voltage relations of Ado and ACh-regulated channels were almost superimposable (Fig. 1B). These channels did not open at random intervals but in bursts. Open time histograms of the channels could be fitted by a single exponential curve with a time constant of about 1.4 msec in both cases

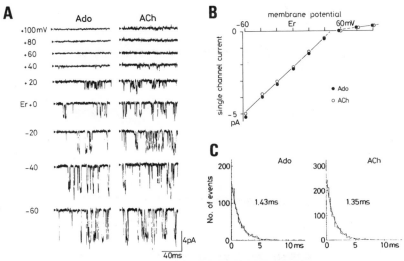

Fig. 1.   Conductance and kinetic properties of Ado and ACh-regulated channel currents. A: examples of Ado and ACh-regulated channel currents from guinea pig atrial cells. The pipette solution contained Ado (10 μM for ACh (5.5 μM) in these experiments. All records were low-pass filtered at 1 kHz (-3dB). The arrowhead by each trace represents the zero current level. B: current-voltage relations of Ado (closed circles) and ACh (open circles) regulated channels. The amplitudes of the channel currents were obtained from amplitude histograms. The slope conductance of the unitary current was 46 pS in this example. C: open time histograms of Ado and ACh-regulated K^+ channels at Er. Both distributions were fitted by a single exponential curve with the time constants indicated in each graph.

(Fig. 1C). We interpret these observations as indicating that ACh and Ado regulate the same K^+ channel in cardiac atrial cell membrane but do so by activating different membrane receptors, *i.e.*, muscarinic ACh receptors and P1-purinergic receptors, respectively (*9*).

III. INVOLVEMENT OF GTP-BINDING PROTEINS IN THE ACTIVATION OF THE K^+ CHANNEL

It was shown that G protein, whose functions are inhibited by IAP is involved in the muscarinic activation of K^+ current in atrial cells of various animals (*6, 7, 17, 18*). We examined effects of intracellular GTP and IAP on activation of the K^+ channel under inside-out patch conditions (*8, 9*).

Figure 2A demonstrates the effects of GTP in the internal side of the membrane on the channel openings regulated by ACh and Ado. The

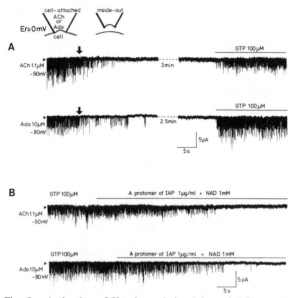

Fig. 2. Activation of K⁺-channels by Ado and ACh requires intracellular GTP and is blocked by IAP. A: the cells were bathed in the internal solution. The concentration of agonists and the patch membrane potential are indicated at each current trace. At the arrow in each trace the patch was excised from the cell, yielding "inside-out" patches. During the period shown by the bar above each current trace, the internal solution containing 100 μM GTP was perfused on the intracellular side of the membrane; channels remained activated. The A promoter of IAP with 1 nM NAD was added to the internal solution containing GTP during the period indicated by the bar above each trace.

cells were bathed in the internal solution. Since Er of the cells in the internal solution could be assumed to be around 0 mV, the patch membrane was hyperpolarized to observe the channel currents. At the arrow in Fig. 2A, the patch membranes were excised from the cells ("inside-out" patch). Under these conditions, the channel openings faded within 1 min in both cases, even though the pipette solution contained agonists. Several minutes after the channel openings had disappeared, GTP (100 μM) was added to the bath solution. The openings of the channel were immediately restored in Ado pipettes as well as in ACh pipettes. The channel openings with GTP were gradually inhibited by preactivated IAP, the A protomer, with NAD (Fig. 2B), indicating that G proteins, G_i and/or G_o, are involved in the activation (19, 20).

If G_i and/or G_o are the common transducers of the signal from muscarinic ACh and Ado receptors to a K⁺ channel, direct activation of

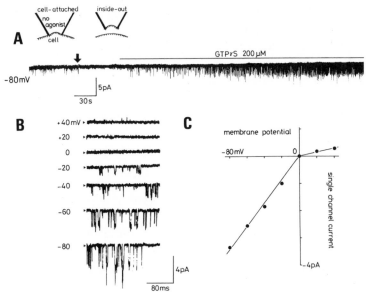

Fig. 3. GTP-γS-induced channels in the inside-out patch membrane. A: effects of GTP-γS applied to the cytoplasmic surface of an excised patch membrane. The pipette solution did not contain agonist. The membrane potential was -80 mV. At the arrow, the inside-out patch was formed. During the period indicated by the bar above the current trace, the internal solution containing 200 μM GTP-γS was perfused. B: examples of GTP-γS-induced channels recorded from the inside-out patch membrane at various membrane potentials. All records were low-pass filtered at 1 kHz. The arrowhead by each trace is the closed level. C: current-voltage relation of the GTP-γS-induced channel. The unitary conductance of inward currents was 43 pS in this example.

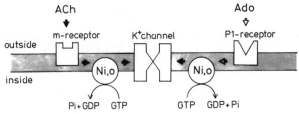

Fig. 4. Simplified scheme of purinergic and muscarinic activation of a K$^+$-channel in atrial cell membrane. In cardiac atrial cell membrane, two different membrane receptors (P1-purinergic and muscarinic ACh) link with K$^+$ channel *via* GTP-binding proteins G_i and/or G_0. This scheme does not represent any quantitative relationships between the two components.

GTP-binding proteins may open the K^+ channel. We examined the effects of GTP-γS, a nonhydrolyzable GTP analogue, on the activation of the channel. In inside-out patches, GTP-γS added to the bath solution caused gradual activation of the same K^+ channel in the absence of agonists in the pipettes (Fig. 3). Therefore, we concluded that muscarinic ACh and Ado receptors are linked to the same K^+ channel *via* GTP-binding proteins, G_i and/or G_o, in the atrial cell membrane (Fig. 4) (*9*).

IV. ROLE OF INTRACELLULAR Mg^{2+} IN THE ACTIVATION OF MUSCARINIC K^+ CHANNEL BY GTP-BINDING PROTEINS

In biochemical studies, it is reported that intracellular Mg^{2+} is essential for activation of G proteins, G_s and/or G_i, which regulate adenylyl cyclase (*21*). To elucidate further the molecular mechanisms underlying activation of the K^+ channel by G proteins, we examined effects of intracellular Mg^{2+} on the K^+ channel activity in the inside-out patch condition (*10*).

Figure 5 shows the effects of intracellular Mg^{2+}, GTP, and GTP-γS on the activation of the K^+ channel. The internal solution contained no Mg^{2+} and 1 mM EDTA instead of EGTA in this series of experiments. The pipette solution contained 1.1 μM ACh or 10 μM Ado. In the

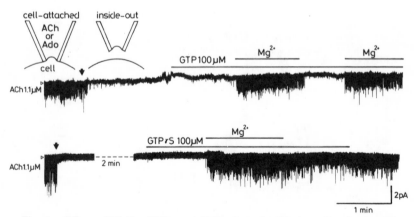

Fig. 5.   Effects of Mg^{2+}-, GTP-, and GTP-γS on the K^+-channel activation. The cells were bathed in the internal solution. The concentration of ACh in the pipettes is indicated at each current trace. The membrane potential of the patch was -60 mV in all cases. At the arrow, the patch was excised from the cell (inside-out patch). The protocol of perfusing Mg^{2+}-, GTP-, and GTP-γS is indicated by the bars above each current trace.

inside-out patch condition, the channel openings disappeared in the absence of the intracellular GTP and Mg^{2+} as observed previously. In the upper current trace of Fig. 5, 100 μM GTP was added to the internal solution 1.5 min after excision of the patch, but openings of the channel did not reappear. When $MgCl_2$ (5 mM) was further added to the solution containing GTP, the channel openings reappeared abruptly. Openings of the channel faded again within 10–20 sec after removal of Mg^{2+} from the solution. These effects of Mg^{2+} in the intracellular side of the membrane were repeatable.

In the lower trace of Fig. 5, we used GTP-γS instead of GTP. Mg^{2+} was also necessary in the internal solution which caused reappearance of the channel openings. However, in this case, once the channels had started to open, the openings persisted, even when both Mg^{2+} and GTP-γS were removed from the internal solution. This observation indicates that activation of the K^+ channel generated by application of GTP-γS is irreversible, which is consistent with the biochemical study showing that activation of G_i by GTP-γS with Mg^{2+} is essentially irreversible (22).

The above observations suggested the following: 1) intracellular

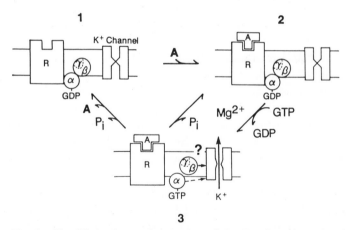

Fig. 6. Simplified scheme of activation of the K^+-channel regulated by GTP-binding proteins in the atrial cell membrane. The meanings of the symbols are as follows: R is a membrane receptor (m-ACh or Ado receptor); A is an agonist (ACh or Ado): α, β, and γ are the subunits of the G protein. In this scheme, α-GTP subunit of G protein was assumed to activate the K^+ channel, but it was also shown that the $\beta\gamma$ subunits can activate the channel. This scheme does not represent any quantitative relationships and does not exclude the possibility of several other steps between each component.

Mg^{2+} is essential for GTP to activate the G protein and 2) deactivation of the G protein may be caused by hydrolysis of GTP to GDP. A simple model for the activation of the K^+ channel by G proteins was proposed earlier (Fig. 6) (10).

V.   EXOGENOUS SUBUNITS OF GTP-BINDING PROTEINS ACTIVATE THE K^+ CHANNEL

The G proteins, which regulate the K^+ channel, are activated by intracellular GTP in a Mg^{2+}-dependent fashion. During activation, it is supposed that the G proteins are functionally dissociated into their subunits, *i.e.*, α-GTP and $\beta\gamma$. Some of the dissociated subunits may be responsible for regulation of the channel openings (10). To identify the subunits which regulate the K^+ channel, effects of various purified subunit proteins from the bovine brain or rat brain were examined in chick embryonic and guinea pig atrial myocytes (12-14, 23).

In chick embryonic atrial cells, IAP-sensitive G proteins also couple m-ACh receptors to a specific inward-rectifying K^+ channel. In inside-out patch conditions, we perfused various kinds of purified subunits of G proteins and found that $\beta\gamma$ subunits were identical; *i.e.*, in both cases the unitary conductance of the inward channel current was about 40–45 pS and the open time histogram of the channel could be fit by a single exponential curve with a time constant of about 1 msec. We interpret the results to indicate that exogenous $\beta\gamma$ subunits can activate the muscarinic K^+ channel in atrial cell membrane.

There has arisen an argument that only α subunits of G proteins activate the channel and that effects of $\beta\gamma$ subunits are due either to the contaminating α subunit or to the detergent, CHAPS, which was used to suspend the $\beta\gamma$ proteins (24-26). Therefore, David Clapham's laboratory and our laboratory decided to reexamine the effects of various subunit proteins on the K^+ channel independently (12, 14, 23, 33).

We studied effects of the $\beta\gamma$ and α-GTP-γS subunits purified from the rat brain on the guinea pig atrial cell K^+ channel (12). The subunit proteins were prepared by Profs. Toshiaki Katada and Michio Ui. We have found that $\beta\gamma$ subunits at a concentration of more than 10 pM activate the K^+ channel in a dose-dependent fashion (Fig. 7). The channel activation saturated > 10 nM and the apparent K_d was 2–3 nM. The detergent CHAPS, which was used to suspend the $\beta\gamma$ protein, never activated the channel between 5 μM and 200 μM; nor did the boiled $\beta\gamma$ preparation activate the channel.

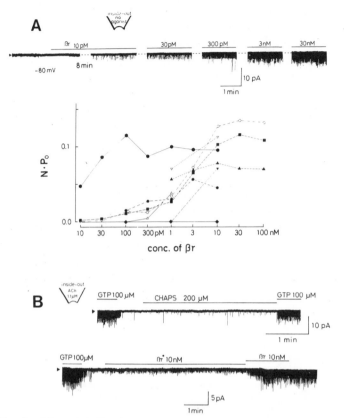

Fig. 7. Muscarinic K^+-channel activation by G protein $\beta\gamma$ subunits. A: inside-out patch recordings from guinea pig isolated atrial myocytes. The $\beta\gamma$ subunits were diluted to 0.01–100 nM in the EGTA-2 mM Mg^{2+} internal solution or in the EGTA-Mg^{2+}-free internal solution and perfused into the internal side of the patch membrane. The lower graph shows the dose-response relation of the $\beta\gamma$ activation of the K^+ channel. The closed symbols represent the relation in EGTA-2 mM Mg^{2+} internal solution. Closed symbols in EGTA-Mg^{2+}-free indicate internal solution. Closed squares in the graph represent the example in the upper current trace. B: detergent, CHAPS, and boiled $\beta\gamma$ subunits did not activate the K^+ channel.

The α subunits (α_{41}-GTPγS, α_{40}-GTPγS, and α_{39}-GTPγS) also activated the K^+ channel but the magnitude of the induced channel openings was much less than those induced by $\beta\gamma$ or GTP-γS (12, 23, 33). The α-activation of the channel could be prevented completely by pretreating the subunits in Mg^{2+}-free EDTA solution containing 20 μM

GDP, while the effects of $\beta\gamma$ subunits were not affected by the same procedure. Therefore, we concluded that the $\beta\gamma$ subunit activation of the K$^+$ channel is specific to the subunit protein (*12*). We believe that the physiological functional arm of G protein to activate the muscarinic K$^+$ channel has not yet been identified (*33*, *34*).

VI.   GTP-BINDING PROTEINS MAY ALSO BE INVOLVED IN THE SHORT-TERM DESENSITIZATION OF ACETYLCHOLINE-INDUCED K$^+$ CURRENT

Desensitization, a tendency of biological responses to wane over time despite the continuous presence of a stimulus, is a widespread physiological process. Although it has been generally believed that the muscarinic response in cardiac preparations shows little desensitization (*4*, *27*), Carmeliet and Mubagwa (*28*) clearly showed in rabbit Purkinje fibers that the ACh-induced increase of cardiac K$^+$ current is desensitized within several minutes after application of ACh. They proposed a three-state receptor model for desensitization, on the assumption that the muscarinic K$^+$ channel is directly associated with muscarinic receptors (*29*). However, as has been shown, the IAP-sensitive G proteins couple

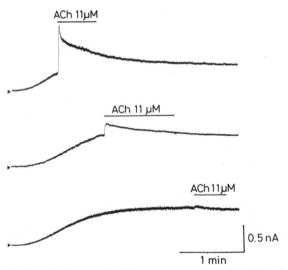

Fig. 8.   Reduction of cellular response to ACh in GTP-γS-loaded cells. The cells were loaded with GTP-γS (100 μM) through the pipettes. The current traces started just after the rupture of the patch membrane. Arrowheads represent the zero current level. Applications of atropine (atr) or ACh are indicated by the bars above the current traces.

m-ACh and Ado receptors to a specific inward-rectifying K^+ channel in atrial myocytes. Thus we characterized the short-term desensitization of the K^+ channel current in single atrial cells loaded with GTP or GTP-γS (11).

In GTP-loaded atrial cells, the ACh-activation of K^+ channel current showed desensitization similar to those reported previously (30). On application of ACh, various amounts of outward current were induced. The maximum current evoked by ACh increased as the concentration of ACh was raised. With ACh higher than 1 μM, the ACh-induced current decreased gradually after having reached a peak. As the ACh concentration was raised, the reduction of the current became more marked and more rapid. In GTP-γS-loaded cells, the muscarinic K^+ channel current increased spontaneously in the absence of agonists, probably due to direct activation of G proteins by GTP-γS (Fig. 8; cells were held at -53 mV). It was found that the cellular response to ACh during the GTP-γS-induced increase of the K^+ current decreased time dependently. In the early phase of the spontaneous increase of the current, ACh accelerated the activation and evoked a large outward current. The ACh-induced

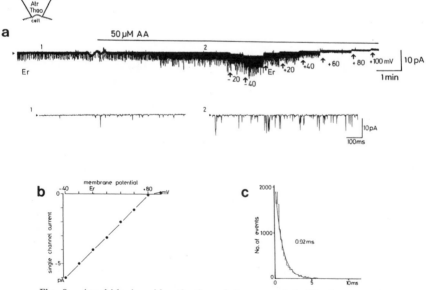

Fig. 9. Arachidonic acid activation of the muscarinic K^+-channel. In the cell-attached patch, when arachidonic acid (AA, 50 μM) was applied to the bath, it activated the muscarinic channel.

current decreased thereafter to a steady level with a time course similar to that observed in GTP-loaded cells (the top current trace in Fig. 8). As the intracellular GTP-γS-induced activation of the current progressed, the magnitude of ACh-induced current transient became smaller and finally negligible (the middle and the bottom current traces in Fig. 8). Similar results were obtained when we used Ado instead of ACh as an agonist.

The above observations indicate that the cellular responses to ACh and Ado both decrease during the GTP-γS-induced activation of the K$^+$ current, without agonist-binding to the receptors. Therefore, we concluded that the short-term desensitization of the muscarinic K$^+$ channel cannot be attributed only to the receptor-ligand interaction and propose the possibility that G proteins may play an essential role in the desensitization process of muscarinic and purinergic regulation of the cardiac K$^+$ channel.

VII. ARACHIDONIC ACID METABOLITES STIMULATE THE G PROTEIN-K$^+$ CHANNEL SYSTEM

We have recently found that arachidonic acid metabolites activate the K$^+$ channel (Fig. 9) (31). Arachidonic acid, applied to the bath solution in the cell-attached form, gradually activated the channel, although the pipette solution contained atropine and theophylline to block the m-ACh and Ado receptors. The arachidonic acid-activation of the channel could be derived from stimulation of GDP-GTP exchange of the G protein caused by the lipoxygenase metabolites of arachidonic acid (31). Since arachidonic acid can be released from the cell membrane by various receptor-dependent stimulations, this novel regulatory pathway of G protein-K$^+$ channel system may play an important role in a variety of physiological and pathophysiological conditions (32).

SUMMARY

Molecular mechanisms underlying activation of cardiac K$^+$ channels by ACh and Ado were examined in isolated atrial myocytes. Electrophysiological studies using patch clamp techniques revealed that GTP-binding proteins, whose functions are inhibited by pertussis toxin, couple muscarinic ACh and Ado receptors to a K$^+$ channel in the cardiac atrial cell membrane. The activation of G proteins by intracellular GTP analogues has an absolute requirement of intracellular Mg^{2+}. During activation, G proteins may be dissociated into their subunits (α-GTP and $\beta\gamma$). Reconstitution experiments using purified subunits from bovine and rat brains showed both subunits can activate the K^+ channel. The maximal effects of the $\beta\gamma$ subunits on the channel activation were more potent than those of α subunits. The $\beta\gamma$-activation of the K^+ channel was due neither to the contamination of activated α subunits in the preparation nor to the detergent (CHAPS) action. A simple model for activation of the channel by G proteins has been proposed. It was shown that G proteins may also be involved in the short-term desensitization of the ACh- or Ado-induced K^+ channel current. The precise mechanism of desensitization is still to be clarified.

Acknowledgments

This work was supported by grants from the Ministry of Education, Science and Culture of Japan, the Research Program on Cell Calcium Signalling in the Cardiovascular System, and by a grant for cardiovascular diseases (1A-1) from the Ministry of Health and Welfare of Japan.

REFERENCES

1. Hutter, O.F. and Trautwein, W. *J. Gen. Physiol.*, **39**, 715 (1956).
2. Trautwein, W. and Dudel, J. *Pflügers Arch.*, **266**, 324 (1958).
3. Noma, A. and Trautwein, W. *Pflügers Arch.*, **377**, 193 (1978).
4. Glitsh, H.G. and Pott, L. *J. Physiol.*, **279**, 655 (1978).
5. Osterrieder, W., Yang, Q.F., and Trautwein, W. *Pflügers Arch.*, **389**, 283 (1981).
6. Pfattinger, P.J., Martin, J.M., Hunter, D.D., Nathanson, N.M., and Hille, B. *Nature*, **317**, 536 (1985).
7. Breitwieser, G.E. and Szabo, G. *Nature*, **317**, 538 (1985).
8. Kurachi, Y., Nakajima, T., and Sugimoto, T. *Am. J. Physiol.*, **251**, H681 (1986).
9. Kurachi, Y., Nakajima, T., and Sugimoto, T. *Pflügers Arch.*, **407**, 264 (1986).
10. Kurachi, Y., Nakajima, T., and Sugimoto, T. *Pflügers Arch.*, **407**, 572 (1986).
11. Kurachi, Y., Nakajima, T., and Sugimoto, T. *Pflügers Arch.*, **410**, 227 (1987).
12. Kurachi, Y., Nakajima, T., Sugimoto, T., Katada, T., and Ui, M. *Pflügers Arch.*, **413**, 325 (1989).
13. Logothetis, D.E., Kurachi, Y., Galper, J., Neer, E.J., and Clapham, D.E. *Nature*, **325**, 321 (1987).
14. Logothetis, D.E., Kim, D., Northup, J.K., Neer, E.J., and Clapham, D.E. *Proc. Natl. Acad. Sci. U.S.A.*, **85**, 5814 (1988).
15. Hamill, O., Marty, A., Neher, E., Sakmann, B., and Sigworth, F.J. *Pflügers Arch.*, **391**, 85 (1981).
16. Belardinelli, L. and Isenberg, G. *Am. J. Physiol.*, **244**, H734 (1983).
17. Endoh, M., Maruyama, M., and Iijima, T. *Am. J. Physiol.*, **249**, H301 (1985).

18. Sorota, S., Tsuji, Y., Tajima, T., and Pappano, A. *Circ. Res.*, **57**, 748 (1985).
19. Florio, V.A. and Sternweis, P.C. *J. Biol. Chem.* **57**, 748 (1985).
20. Katada, T. and Ui, M. *Proc. Natl. Acad. Sci. U.S.A.*, **79**, 3129 (1982).
21. Bokoch, G.M., Katada, T., Northup, J.K., Hewlett, E.L., and Gilman, A.G. *J. Biol. Chem.*, **258**, 2072 (1983).
22. Katada, T., Bokoch, G.M., Northup, J.K., Ui, M., and Gilman, A.G. *J. Biol. Chem.* **259**, 3568 (1984).
23. Kobayashi, I., Shibasaki, H., Takahashi, K., Tohyama, K., Kurachi, Y., Ito, H., Ui, M., and Katada, T. *Eur. J. Biochem.*, **191**, 499 (1990).
24. Yatani, A., Codina, J., Brown, A.M., and Birnbaumer, L. *Science*, **235**, 207 (1987).
25. Codina, J., Yatani, A., Grenet, D., Brown, A.M., and Birnbaumer, L. *Science*, **236**, 442 (1987).
26. Kirsch, G.E., Yatani, A., Codina, J., Birnbaumer, L., and Brown, A.M. *Am. J. Physiol.*, H1200 (1988).
27. Noma, A., Pepper, K., and Trautwein, W. *Pflügers Arch.*, **381**, 255 (1979).
28. Carmeliet, E. and Mubagwa, K. *J. Physiol.*, **371**, 239 (1986).
29. Soejima, M. and Noma, A. *Pflügers Arch.*, **400**, 421 (1984).
30. Nilius, B. *Biomed. Biochim. Acta*, **42**, 519 (1983).
31. Kurachi, Y., Ito, H., Sugimoto, T., Shimizu, T., Miki, I., and Ui, M. *Nature*, **337**, 555 (1989).
32. Kurachi, Y., Ito, H., Sugimoto, T., Shimizu, T., Miki, I., and Ui, M. *Pflügers Arch.*, **414**, 102 (1989).
33. Navarati, C., Clapham, D.E., Ito, H., and Kurachi, Y. *In* "G Proteins and Signal Transduction," ed. N.M. Nathanson and T.K. Harden, p. 29 (1990). The Rockefeller Univ. Press, New York.
34. Kurachi, Y. *NIPS*, **4**, 158 (1989).

Regulation of Intracellular Ca^{2+} Transients of Myocardial Cell

MASAO ENDOH

Department of Pharmacology, Yamagata University School of Medicine, Yamagata 990-23, Japan

Calcium ions play a central role in the coupling of membrane excitation to initiation of development of the contractile force (E-C coupling) by binding to troponin C, resulting in disinhibition of troponin I. It is, therefore, critical to assess the relation between changes in the intracellular Ca^{2+} concentration ($[Ca^{2+}]_i$) and the force of contraction simultaneously, in order to analyze the subcellular regulatory mechanism of cardiac E-C coupling. Regulation of the myocardial contractility induced by divergent physiologic and pathophysiologic interventions is achieved either by modulation of the intracellular Ca^{2+} mobilization, modulation of the Ca^{2+} sensitivity of myofibrils or both. Various cellular and intracellular structures are involved in the regulation of cardiac E-C coupling: sarcolemmal receptor and ion transport system (channels and exchangers), sarcoplasmic reticulum membrane Ca^{2+} pump and Ca^{2+} release channels, contractile proteins and some other structures such as mitochondria and Ca^{2+} binding cytosolic proteins (Fig. 1). The role of Ca^{2+} in cardiac E-C-coupling can be deduced from modulation of the contractile function together with biochemical findings in *in vitro* studies such as $^{45}Ca^{2+}$ uptake by myocardial cells or isolated membrane vesicles derived from sarcoplasmic reticulum, and activation of actomyosin ATPase or skinned cardiac fibers by Ca^{2+}.

In 1978, it was first demonstrated that application of aequorin extracted from jellyfish *Aequorea aequorea* (or *Aequorea folskarea*) to an isolated intact frog heart preparation made it possible to analyze the

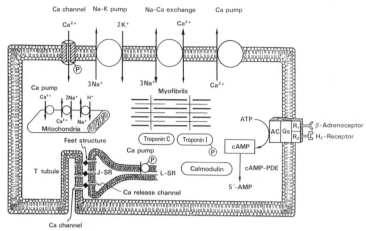

Fig. 1. Schematic representation of cellular organelles involved in cardiac E-C coupling.

relation between changes in $[Ca^{2+}]_i$ and force of contraction in intact myocardial cells (1). Since then, this experimental procedure has been applied to mammalian atrial and ventricular muscle isolated from various species, and has been demonstrated to be of extreme usefulness for analyses of the E-C coupling process in myocardial cells (2–4). The present paper describes briefly the experimental procedure of applying aequorin to mammalian cardiac muscle and the findings obtained. As examples of the physiologic regulation, influence of alteration of the extracellular Ca^{2+} concentration ($[Ca^{2+}]_o$), changes in muscle-length, frequency of stimulation, temperature, and the effects of sympathomimetic agents and some cardiotonic agents will be described. The changes during ischemia, hypoxia, Ca^{2+} paradox, and heart failure will likewise be briefly discussed based on recent findings using similar experimental methods (28–30).

I. APPLICATION OF AEQUORIN TO THE CARDIAC MUSCLE

    Hearts excised from various mammalian species including rabbits, rats, dogs, cats, ferrets, and guinea pigs following anesthesia and exsanguination can be used for the experiments. Papillary muscles, free-running trabeculae or atrial muscles are dissected from the ventricle or atrium. Aequorin is microinjected into the superficial cells of the muscle mounted horizontally in an organ bath constructed for aequorin injection. Bicarbonate-buffered Krebs-Henseleit solution freshly oxygenated with 95% O_2 and 5% CO_2 is circulated rapidly through the surface of muscle at 32°C. The muscle is stimulated at a rate of 0.5 Hz with threshold pulses delivered through punctate electrodes and stretched to a length at which contractile force is nearly maximum. After an equilibration period of about 30 min, electrical stimulation is discontinued and aequorin microinjection is started. Aequorin is dissolved at a concentration of about 2 mg/ml in a solution containing 150 mM KCl and 5 mM HEPES (N-2-hydroxyethylpiperazine-N'-2-ethanesulfonic acid) buffer, pH 7.5. The solution is loaded into fine-tipped (35–50 megohms in 150 mM KCl) micropipettes through which membrane potential is monitored to determine when cells have been penetrated. Aequorin is injected by the application of gas pressure (1–4). When a satisfactory light signal is obtained on stimulation (after injection of aequorin solution into 50–100 superficial cells), the muscle is transferred to an apparatus designed to record light signals with high efficiency and to minimize motion artifacts in the aequorin signals, which is a modified version of the original construction by Blinks et $al.$ (4). The muscle is stimulated with pulses of 5-msec duration and about 20% above threshold intensity at a rate of 0.5–1.0 Hz through the punctate cathode. The top of the muscle is connected with 9-0 Tevdek thread to the arm of a servo-operated electromagnet muscle lever (5) operated in the isometric mode. During a 30 min equilibration period after the transfer, the muscle length is adjusted to that at which contractile force is maximum. Experiments are usually carried out at 37.5°C. The light emitted by the injected aequorin is detected by a photomultiplier (EMI 9635A). Signal-averaging is carried out to obtain a satisfactory signal-to-noise ratio in aequorin signals. In the records shown in the present study, between 32 and 256 signals are averaged. The preparations are usually exposed to ($\pm$)-bupranolol (3×10^{-7} M) for more than 20 min before experimental interventions in order to avoid modulation by norepinephrine released in the changes induced by specific interventions, except β-adrenoceptor agonists.

II. ALTERATION OF $[Ca^{2+}]_o$

Amplitudes of Ca^{2+} transients as well as isometric contractions change depending on the $[Ca^{2+}]_o$ substantially in parallel. The relation between the changes in Ca^{2+} transients (the power of -2.5 of the height of aequorin signals) and the force is on a straight line (6). Since the time courses of the two signals are not modified by alteration of $[Ca^{2+}]_o$, it is supposed that the modulation induced by this manipulation may be the simplest among the various interventions. Therefore, influence of the alteration of $[Ca^{2+}]_o$ can be employed as the standard by which to evaluate the mode of regulation by a variety of interventions.

III. LENGTH-TENSION RELATIONSHIP

Increase in muscle length elevates the Ca^{2+} sensitivity of skinned cardiac muscle preparations (7, 8). The findings in an aequorin-injected guinea pig atrial trabecula indicate that a similar mechanism operates in the intact myocardial cell preparation. As shown in Fig. 2, upper panel, when the muscle length is varied in a wide range between the slack and L_{max}, the isometric contraction changes in a graded manner depending on the length of the muscle. The Ca^{2+} transients change very little and are of essentially the same amplitude at various muscle lengths, as can be seen from tracings at the shortest, middle, and longest muscle length superimposed (Fig. 2, lower panel). Close inspection of the superimposed tracings reveals that the Ca^{2+} transient recorded at L_{max} declines slightly earlier than the one recorded at slack length. Similar results were obtained in each of seven rabbit papillary muscles and one rabbit atrial trabecula studied at 37.5°C. In cat and rat papillary muscles studied at 30°C, a more pronounced stretch-induced abbreviation of aequorin signals has been reported (9). It remains to be determined whether the quantitative discrepancies between these experimental findings are due to a species difference, a difference in experimental conditions or both. In the preparations that we studied at 37.5°C, there was little or no time-dependent effect of a change in muscle length on the amplitude of the aequorin signal (although a dramatic effect on the strength of contraction had been induced). On the other hand, a gradual secondary (time-dependent) change in the strength of contraction has been shown to take place under some experimental conditions, especially at relatively low experimental temperatures (9- 12). The mechanism of this gradual change after alteration of muscle length has not been established, and there is conflicting evidence about the involvement of the potential mechanism such as release of norepinephrine (10).

IV. FREQUENCY-INDUCED REGULATION

Amplitude of Ca^{2+} transients and strength of isometric contractions

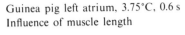

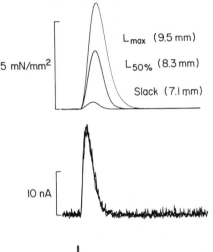

Fig. 2.   Influence of alteration of muscle length on Ca²⁺ transients (lower panel) and isometric contractions (upper panel) of an aequorin-injected strip of guinea pig left atrium (muscle weight, 4.4 mg). The strip was stretched from slack length to L_{max} (10 mm). The aequorin signals were of essentially the same amplitude at all muscle lengths (those at slack length, 50% L_{max} and L_{max} are superimposed at the bottom). Signal averaging of 64 successive contractions was carried out.

are affected markedly by altering the frequency of stimulation. When the stimulus interval is prolonged in a graded manner, the peak of Ca²⁺ transients and isometric contractions decreases gradually depending on the stimulus interval (Fig. 3). Both Ca²⁺ transients and isometric contractions were little affected by prolonging the stimulus interval from 16 to 32 sec and longer, indicating that the rested state contraction is achieved over the range of stimulus intervals.

At the interval between 16 and 2 sec, strength of isometric contractions is definitely increased by the abbreviation of intervals with relatively small changes in the amplitude of Ca²⁺ transients. The duration of Ca²⁺ transients apparently increased with an abbreviation of the interval of this range.

When the stimulus interval is abbreviated further from 2 to 1 sec and shorter up to 0.35 sec, the peak of Ca²⁺ transients and isometric contractions increases essentially in parallel. At this range of stimulus

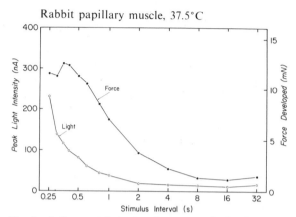

Rabbit papillary muscle, 37.5°C

Fig. 3. Influence of altering frequency of stimulation on Ca²⁺ transients and isometric contractions of an aequorin-injected rabbit papillary muscle (length, 7.5 mm; cross-sectional area, 0.85 mm²). Signal averaging was done between 32 to 256 successive contractions.

intervals, the duration of Ca²⁺ transients and isometric contractions is shortened by an abbreviation of the interval.

When the interval is shortened further, the peak of Ca²⁺ transients increases further, whereas the strength of isometric contractions becomes steady or is decreased slightly after having reached the maximum at the interval of 0.4 or 0.35 sec.

Thus, while the changes in contractile force induced by alteration of the frequency are essentially due to the modulation of Ca²⁺ mobilization, close inspection of the influence of frequency over a wide range indicates that the frequency induces more complex regulation: not only the peak, but also the duration of Ca²⁺ transients, and the relation between the Ca²⁺ transients and force are modified depending on the range of frequency.

The amplitude and duration of the Ca²⁺ transient detected by application of aequorin represents the changes in Ca²⁺ concentrations in the cytoplasm that are determined by the balance between the rate at which Ca²⁺ enters the cytoplasm and the rate at which it is bound or sequestered. Troponin C and sarcoplasmic reticulum are quantitatively the most important Ca²⁺ store sites in the myocardial cells. Troponin C binds Ca²⁺ so rapidly that only a small fraction of Ca²⁺ entering the cytoplasm may be free to bind aequorin molecules in the cytoplasm to yield the Ca²⁺ transient. While the outward transport of Ca²⁺ from the myocardial cells through the sarcolemma either by Na^+-Ca^{2+} exchanger or by Ca^{2+} pump may play a role in regulation of the duration of Ca^{2+} transients, the primary determinant of this duration is considered to be the uptake of Ca^{2+} by the sarcoplasmic reticulum. During the early part of the declining phase of aequorin signals troponin C may release an amount of Ca^{2+} into the cytoplasm, and thereby slow the rate of decline and prolong the duration of Ca^{2+} transients. The β-adrenoceptor stimulation and the elevation of $[Ca^{2+}]_o$ affect the early part of the declining phase of aequorin signals in a similar manner, when the force of contraction is increased by both interventions to the same extent (13).

The rate of Ca^{2+} uptake by the sarcoplasmic reticulum is regulated by phospholamban, the functions of which are regulated by the extent of phosphorylation produced either by protein kinase A activation or the calmodulin-dependent process. Evidence that the former plays an important role will be discussed later. The activity of the latter in regulation of the myocardial contractile force has not yet been established.

At stimulus intervals of 2 sec and longer (at frequencies of 0.5 Hz and lower) prolongation of Ca^{2+} transients with little change in the amplitude of the Ca^{2+} transient appears to be responsible for the Bowditch positive staircase phenomenon. This may be due to the imbalance between the Ca^{2+} release from troponin C and the rate of Ca^{2+} uptake by the sarcoplasmic reticulum. The findings that W-7 and trifluoperazine [calmodulin antagonists *in vitro* (14)] did not affect the Ca^{2+} transient and force at this range of frequency (though they effectively inhibited these parameters at higher frequencies) imply that the calmodulin-mediated mechanism may be less significant at low frequencies of stimulation (15).

The positive staircase phenomenon at the range of stimulus intervals between 2 and 0.4 sec (0.5–2.5 Hz) is readily explainable based on the mechanism accepted generally as the cardiac E-C coupling process: an increased Ca^{2+} influx through voltage-dependent calcium channels which are activated more effectively by increasing frequencies of repetitive depolarization, and the resultant Ca^{2+} loading to the sarcoplasmic reticulum; accumulation of the intracellular Na^+, resulting in modulation of Na^+-Ca^{2+} exchange system.

The present findings indicate also that an elevation of frequency to a rate higher than 2.5 Hz further facilitates the Ca^{2+} mobilization to a level much higher than that required to activate the myofibrils to the maximum extent. β-Adrenoceptor activation is able to elevate the amplitude of Ca²⁺ transients to a level even higher than that reached by raising the frequency (*13*).

V. INFLUENCE OF TEMPERATURE

The amplitude of twitch contractions is known to be increased prominently in association with prolongation of contraction duration by lowering the experimental temperature. Influence of the lowering of temperature from 37.5°C to 32.5°C on the Ca²⁺ transient and isometric contraction is shown in Fig. 4. The amplitude and duration of both Ca²⁺ transients and isometric contractions increase substantially in parallel with lowering temperature. It is considered that the energy requiring process of E-C coupling (mediated by sarcolemmal and sarcoplasmic reticulum membrane ion pump ATPases) may be influenced most readily by changing the temperature. The time to peak of both parameters increased in parallel, while the rate of decline of the Ca²⁺ transient and isometric contraction was little influenced by lowering the temperature. From tracings in Fig. 4, it appears that the relation between decline of the Ca²⁺ transient and relaxation of contraction is straightforward. However, it is noteworthy that the decreased activity of the sarcoplasmic reticulum Ca²⁺ pump caused by lowering the temperature reflects on the prolongation of the time to peak response of both parameters but not on the rate

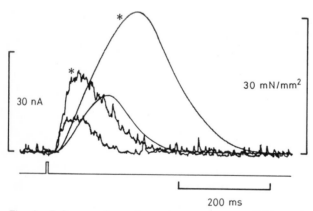

Fig. 4.   Influence of experimental temperature on Ca²⁺ transients and isometric contractions of an aequorin-injected rabbit papillary muscle (length, 7.5 mm; cross-sectional area, 0.85 mm²). Signal averaging of 128 successive contractions was carried out. The temperature was lowered from 37.5°C to 32.5°C (records with asterisks). Stimulus interval: 1 sec.

of relaxation. It is still unclear what the essential process is regulating the relaxation of cardiac muscle. It is generally believed that in cardiac muscle contracting under the physiological condition the rate of Ca^{2+} uptake by the sarcoplasmic reticulum Ca^{2+} pump is a critical determinant of relaxation, as described in the previous section; during the induction of the positive inotropic effect of β-adrenoceptor agonists, both the decrease of Ca^{2+} sensitivity of troponin C in relation to phosphorylation of troponin I, and the increased rate of Ca^{2+} uptake by phosphorylation of phospholamban may play an important role. Relation of the decline of Ca^{2+} transients and that of isometric contractions is unknown. Yue (16) has concluded that the rate of decline of Ca^{2+} transients reflects on the prolongation of time to peak force but not on the rate of relaxation of contraction, based on the computer analysis of aequorin signals and associated contractions.

Relation between changes in the amplitude of Ca^{2+} transients and isometric contractions by the alteration of temperature was investigated at different levels of the force of contraction induced by changing the frequency of stimulation. The relation was shifted slightly to the left and upward probably because of the prolongation of Ca^{2+} transients. Changes in the duration of Ca^{2+} transients affect the relation since the equilibrium between Ca^{2+} transients and force is not achieved during twitch contractions (16).

It has to be taken into consideration that this change of temperature may have modified the relation between the Ca^{2+} concentration and aequorin bioluminescence. In preliminary experiments, however, it was found that the relation was scarcely modified by the difference of temperature in this range (Blinks, personal communication).

VI. EFFECTS OF SYMPATHOMIMETIC AMINES

Effects of sympathomimetic amines including catecholamines (isoproterenol, norepinephrine, epinephrine, and dopamine) and phenylephrine on the relation of Ca^{2+} transients and isometric contractions have been analyzed in detail in the aequorin-injected mammalian ventricular myocardium including rabbit (13), rat (17), and ferret (18) papillary muscles. Although there exist quantitative differences in the response to sympathomimetic amines among species, characteristics of the catecholamine-induced regulation are essentially similar. In brief, sympathomimetic amines promote mobilization of the intracellular Ca^{2+} through activation of β- and/or α-adrenoceptors. α-Adrenoceptor stimulation is much less effective than β-adrenoceptor stimulation (*13*). β-Adrenoceptor stimulation decreases (*13–15*), while α-stimulation may increase the Ca^{2+} sensitivity of contractile proteins (*13*).

VII. EFFECTS OF CARDIOTONIC AGENTS

Cardiotonic agents are classified largely to two classes: one, exerting the effect through the cellular cyclic AMP-dependent process and the other, through the cyclic AMP-independent process. The representative agent of the latter is cardiac glycoside digitalis. Most of the pre-existing and newly developed cardiotonic agents belong to the former: theophylline, glucagon, histamine (H_2-receptor-mediated effect), forskolin, and new positive inotropic agents (amrinone, milrinone, enoximone, piroximone, vesnarinone, sulmazole, MCI-154, *etc.*). Those agents belonging to the former elicit the positive inotropic effect resulting in changes similar to those of sympathomimetic amines mediated by β-adrenoceptors.

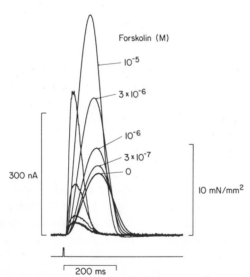

Rabbit papillary muscle, 37.5°C, 1 s
($\pm$)-Bupranolol 3×10^{-7} M

Fig. 5.   Concentration-dependent effects of forskolin on Ca^{2+} transients and isometric contractions of an aequorin-injected rabbit papillary muscle (length, 7.0 mm; cross-sectional area, 0.43 mm²). Signal averaging was made of 128 successive contractions.

Figure 5 shows the effect of forskolin on the rabbit papillary muscle. The amplitude of Ca^{2+} transients and isometric contractions is facilitated markedly in a concentration-dependent manner with an abbreviation of the duration of both signals. An acceleration of the relaxation is more pronounced.

Among these agents belonging to the former category, certain agents (theophylline, sulmazole, MCI-154, and vesnarinone, but to a lesser extent) produce a dissociation of the force from Ca^{2+} transients in a direction indicating an elevation of Ca^{2+} sensitivity of contractile proteins in aequorin-injected intact cardiac cell preparations (19–22). These findings are consistent with the observations in *in vitro* experiments employing actomyosin ATPase and skinned cardiac fibers (23, 24) that the sensitivity to Ca^{2+} is increased by these cardiotonic agents. The physiologic relevance of these additional actions in regulation of the cardiac contractility in the treatment of patients with myocardial failure is not yet determined unequivocally as long as concentrations higher than those required for cyclic AMP phosphodiesterase inhibition are necessary in animal experiments.

VIII. CHANGES DURING Ca^{2+} PARADOX

After exposure of cardiac tissues to Ca^{2+} free solution, myocardial cells become permeable to Ca^{2+} and are overloaded with Ca^{2+} when the physiological solution with control $[Ca^{2+}]_o$ is reintroduced (25–27). While it is known that the contractile function is dissociated from cellular Ca^{2+} accumulation, it remains to be established whether the Ca^{2+} transient is depressed (even though the overall cellular calcium content is increased), or uncoupling of the contractile activation from the Ca^{2+} transient occurs.

This series of experiments was carried out to clarify this point. Changes in Ca^{2+} transients and isometric contractions during exposure to nominally Ca^{2+} free solution and readministration of Ca^{2+} are presented in Fig. 6. During Ca^{2+} free perfusion, Ca^{2+} transients and isometric contractions decreased gradually almost to the resting levels. When $[Ca^{2+}]_o$ is elevated from nominally free to 1/8 (d) and 1/4 (e) of that in the standard solution (2.5 mM), the Ca^{2+} transients are increased prominently to a level much higher than the control, in association with after-glimmers, but the force of contraction recovers only by 50% of the control. Thus, there occurs a marked dissociation of force from Ca^{2+}

A. Ca²⁺ free

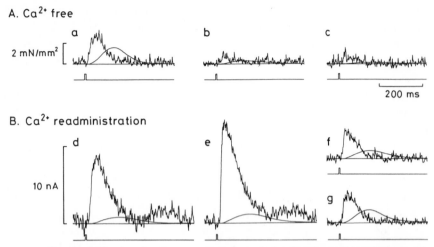

Fig. 6. Influence of exposure to Ca²⁺ free solution for 15 min (b) and 30 min (c), and of readministration of Ca²⁺ to raise the $[Ca^{2+}]_o$ to 1/8 (d), 1/4 (e), 1/2 (f) of normal solution (2.5 mM) and normal (g) on Ca²⁺ transients (noisy tracings) and isometric contractions in an aequorin-injected rabbit papillary muscle (length, 5.5 mm; cross-sectional area, 0.65 mm²). Signal averaging of 32 successive contractions was carried out.

transients. When the $[Ca^{2+}]_o$ is elevated further to 1/2 (f) and to the normal level (g), Ca²⁺ transients become smaller and return to a level near the control, the force of contraction increasing gradually to reach the control level. The resting level of $[Ca^{2+}]_i$ is also elevated as can be seen from the increased noise level during the readministration of Ca²⁺.

When $[Ca^{2+}]_o$ is elevated further to ×2, ×4, and ×6 of the standard solution, the Ca²⁺ transients and isometric contractions increase as usually observed in the rabbit papillary muscle without exposure to the Ca²⁺ free solution. Dissociation of isometric contractions from the peak of Ca²⁺ transients during readministration of Ca²⁺ is obvious. Both parameters recover completely to the control levels prior to Ca²⁺ free exposure, but during the recovery the relation between the peak of Ca²⁺ transients and isometric contractions markedly deviates from that observed by alteration of $[Ca^{2+}]_o$ without exposing to the Ca²⁺ free solution, indicating the uncoupling of activation of contractile proteins from the Ca²⁺ transient (Fig. 6). Similar experiments have been carried out in nine isolated rabbit papillary muscles. An increase in the peak of Ca²⁺ transients with after-glimmers, and the dissociation of contractile force from the enhanced Ca^{2+} transient are consistently observed in these muscles.

The present observations indicate that the changes in Ca^{2+} signaling during Ca^{2+} paradox involve both an abnormality of Ca^{2+} mobilization and uncoupling of contractile proteins from intracellular Ca^{2+} signals. It is noteworthy that these changes in E-C coupling during Ca^{2+} paradox under the present experimental conditions (gradual elevation of $[Ca^{2+}]_o$ following the 30 min exposure to Ca^{2+} free solution) are apparently reversible. In this context, further studies will be required to elucidate whether these changes in the present experiments can be extended to the more severe Ca^{2+} paradox or whether a qualitatively different mechanism may participate contributing to the changes during more severe Ca^{2+} paradox.

IX. INFLUENCE OF HYPOXIA, ISCHEMIA, AND HEART FAILURE

The regulation of myocardial contractility in relation to changes in Ca^{2+} transients produced by pathophysiologic interventions in the intact myocardial cells is currently attracting the keen interest of most researchers in the cardiovascular research field, but largely remains to be solved in future studies. Findings in the human ventricular myocardial tissues isolated from patients with dilated and hypertrophic cardiomyopathy (loaded with aequorin by chemical loading procedures) show clearly that the process of Ca^{2+} mobilization (especially disturbance of the Ca^{2+} uptake process) may be primarily involved in the pathologic changes in Ca^{2+} signaling in heart failure patients (28). On the other hand, the sensitivity of contractile proteins to Ca^{2+} does not appear to play an essential role (29). Modulation of Ca^{2+} signaling during hypoxia and ischemia has been elegantly demonstrated in recent experiments using aequorin-loaded perfused heart preparations (30).

SUMMARY

Changes in $[Ca^{2+}]_i$ during cardiac muscle contractions can be assessed by application of the Ca^{2+} sensitive bioluminescent protein aequorin to the isolated multicellular intact cardiac muscle preparations. The modulation of cardiac E-C coupling by various physiologic and pathophysiologic interventions is summarized in Table I. Regulation by these interventions is achieved by modulation of intracellular Ca^{2+}

TABLE I

Modulation of Ca^{2+} Mobilization and Ca^{2+}-Force Relation by Physiologic and Pathophysiologic Regulatory Interventions

	Ca^{2+} mobilization	Ca^{2+}-force relation
I. Physiological interventions		
$[Ca^{2+}]_o$	+++	No
Length tension relation	No	+ (coupling)
Frequency-force relation	+++	No [(−) saturation]
Temperature	++	No [(+) duration]
Adrenoceptor activation		
β-Adrenoceptors	+++	− (Ca^{2+} sensitivity)
α-Adrenoceptors	+	+ (Ca^{2+} sensitivity)
Cardiotonic agents		
Amrinone, milrinone	++	− (Ca^{2+} sensitivity)
Sulmazole, theophylline	++	+ (Ca^{2+} sensitivity)
II. Pathophysiologic interventions		
Ca^{2+} paradox	++	− (uncoupling)
Heart failure (28, 29)	[(+) duration]	No
Hypoxia (30)	++	No
Ischemia (30)	++	− (uncoupling)

+: facilitation or increase; −: inhibition or decrease; no: no effect.

mobilization and/or modulation of Ca^{2+} sensitivity of myofibrils. The simplest regulation may be achieved by changing $[Ca^{2+}]_o$, and this may be employed as a standard to evaluate the changes in Ca^{2+}-force relation elicited by various regulatory interventions. Ca^{2+} transients change little during the length-dependent regulation of contractility, indicating that this regulation is achieved essentially by modulation of the Ca^{2+} sensitivity of contractile proteins. Regulation of contractility by changes in frequency is mainly achieved by modulation of Ca^{2+} mobilization, while additional mechanisms are involved depending on the range of frequency of stimulation. Sympathomimetic amines promote the Ca^{2+} mobilization through activation of β- and/or α-adrenoceptors, α-adrenoceptor stimulation being much less effective than β-stimulation in this respect. β-Adrenoceptor stimulation decreases, while α-stimulation may increase the Ca^{2+} sensitivity of contractile proteins. Certain cardiotonic agents such as sulmazole, theophylline, MCI-154, and vesnarinone cause a dissociation of contractile force from the amplitude of Ca^{2+} transients, indicating that these agents may increase the Ca^{2+} sensitivity of contractile proteins in relatively higher concentrations. Thus, the aequorin-injected multicellular intact myocardial cell preparation provides an excellent experimental procedure to address the physiologic and pharmacologic modulation of E-C coupling in the mammalian myocardium.

Elucidation of details of the subcellular mechanism involved in the pathophysiologic modulation of Ca^{2+} signaling in myocardial cells awaits further study.

Acknowledgments

I wish to acknowledge the continuous encouragement and help of Dr. J. R. Blinks, Department of Pharmacology, Mayo Clinic, Rochester, Minnesota throughout the course of this work. This work was partly supported by Grant-in-Aid (No. 62624004) for Scientific Research on Priority Areas, from The Ministry of Education, Science and Culture, Japan and National Institutes of Health grant HL-12186.

REFERENCES

1. Allen, D.G. and Blinks, J.R. *Nature*, **273**, 509 (1978).
2. Blinks, J.R. *In* "Methods for Studying Heart Membranes," Vol. II, ed. N.S. Dhalla, p. 237 (1984). CRC Press, Boca Raton.
3. Blinks, J.R., Mattingly, P.H., Jewell, B.R., Van Leeuwen, M., Harrer, G.C., and Allen, D.G. *Methods Enzymol.*, **57**, 292 (1978).
4. Blinks, J.R., Wier, W.G., Hess, P., and Prendergast, F.G. *Prog. Biophys. Mol. Biol.*, **40**, 1 (1982).
5. Brutsaert, D.L. and Claes, V.A. *Circ. Res.*, **35**, 345 (1974).
6. Endoh, M. *In* "New Aspects of the Role of Adrenoceptors in the Cardiovascular System," ed. H. Grobecker, A. Philippu, and K. Starke, p. 78 (1986). Springer-Verlag, Berlin, Heidelberg.
7. Fabiato, A. and Fabiato, F. *J. Gen. Physiol.*, **72**, 667 (1978).
8. Hibberd, M.G. and Jewell, B.R. *J. Physiol.*, **329**, 527 (1982).
9. Allen, D.G. and Kurihara, S. *Eur. Heart J.*, **1** (Suppl. A), 5 (1980).
10. Parmley, W.W. and Chuck, L. *Am. J. Physiol.*, **224**, 1195 (1973).
11. Gulch, R.W. and Jacob, R. *Pflügers Arch.*, **357**, 335 (1975).
12. Lakatta, E.G. and Jewell, B.R. *Circ. Res.*, **40**, 251 (1977).
13. Endoh, M. and Blinks, J.R. *Circ. Res.*, **62**, 247 (1988).
14. Hidaka, H., Yamaki, T., Naka, M., Tanaka, T., Hayashi, H., and Kobayashi, R. *Mol. Pharmacol.*. **17**, 66 (1980).
15. Endoh, M. *In* "Advances in Experimental Medicine and Biology: Calcium Protein Signaling," Vol. 255, ed. H. Hidaka, E. Carafoli, A.R. Means, and T. Tanaka, p. 461 (1989). Plenum Press, New York, London.
16. Yue, D.T. *Am. J. Physiol.*, **252**, H760 (1987).
17. Kurihara, S. and Konishi, M. *Pflügers Arch.*, **409**, 427 (1987).
18. Okazaki, O., Suda, N., Hongo, K., Konishi, M., and Kurihara, S. *J. Physiol.*, **423**, 221 (1990).
19. Blinks, J.R. and Endoh, M. *J. Physiol.*, **353**, 63P (1984).
20. Blinks, J.R. and Endoh, M. *Circulation*, **73** (Suppl. III), III-85 (1986).
21. Endoh, M., Yanagisawa, T., Taira, N., and Blinks, J.R. *Circulation*, **73** (Suppl. III),

III-117 (1986).

22. Endoh, M. *In* "Recent Advances in Calcium Channels and Calcium Antagonists," ed. K. Yamada and S. Shibata, p. 51 (1990). Pergamon Press, New York.
23. Solaro, R.J. and Rüegg, J.C. *Circ. Res.*, **51**, 290 (1982).
24. Kitada, Y., Narimatsu, A., Matsumura, N., and Endo, M. *J. Pharmacol. Exp. Ther.*, **243**, 633 (1987).
25. Crevey, B.J., Langer, G.A., and Frank, J.S. *J. Mol. Cell. Cardiol.*, **10**, 1081 (1978).
26. Nayler, W.G., Perry, S.E., Elz, J.S. and Daly, M.J. *Circ. Res.*, **55**, 227 (1984).
27. Øksendal. A.N., Jynge, P., Sellevold, O.F.M., Rotevatn, S., and Saetersdal. T. *J. Mol. Cell. Cardiol.*, **17**, 959 (1985).
28. Gwathmey, J.K., Copelas, L., MacKinnon, R., Schoen, F.J., Feldman, M.D., Grossman, W., and Morgan, J.P. *Circ. Res.*, **61**, 70 (1987).
29. Gwathmey, J.K. and Morgan, J.P. *Circ. Res.*, **57**, 836 (1985).
30. Kihara, Y., Grossman, W., and Morgan, J.P. *Circ. Res.*, **65**, 1029 (1989).

Subject Index

222

DATE DUE

AP 8'96			